I0558420

Permaculture Organic Gardening

for beginners

ELISABETH FEKONIA

Copyright 2023 by Elisabeth Fekonia

All rights reserved. This book or any portion thereof may not
be reproduced or used in any manner whatsoever without
the express written permission of the publisher except for
the use of brief quotation in a book review.

Inquiries and Book Orders should be addressed to:

 **Great Writers Media**

Great Writers Media
Email: info@greatwritersmedia.com
Phone: 877-556-0487

ISBN: 979-8-89175-042-5 (sc)

FOREWORD

My passion for vegetable gardening goes right back to when we first immigrated to Australia from Holland. I was nine years old, and my father decided to dig over the entire backyard and shape the soil into garden beds. We bought some seeds and off we went into the exciting world of vegetable gardening. Coming from a cold climate and from high rise housing, we had no clue how to go about it, but luckily for us, the seeds came up and we gradually had a vegetable garden.

That was entirely built on good luck as the soil must have been decent enough to sustain growth and produce a yield. It was my introduction to gardening, but I certainly had to learn a few things along the way! Wherever I lived I had a vegie garden. It was something that always stayed with me over the years and now well into my sixties I'm as excited as ever to see my gardens grow.

When I was introduced to Permaculture, I took to it like a duck to water. Permaculture was a language I understood, it gave me plenty of thought on how to have better yields and efficiency. I fell in love with the whole concept of planning, designing, guild planting, water harvesting, composting, worm farming and so much more. My gardening life has become so much richer and satisfying as Permaculture has given me many tools to work with.

Over the years my gardening methods have become my own as I adapted to the vagaries of my land and soil. Every garden is different with its own set of challenges but with perma knowledge also comes a laid back and relaxed attitude as I know I don't always have to follow standard gardening practices. This book has been designed to be a quick easy read with one clear message. When in doubt, add organic matter! For soil fertility add organic matter and for moisture conservation add organic matter, for pest control add organic matter. Do you get the idea?

Many gardening books will tell you to practice rotational gardening, companion planting and a myriad of other techniques but I have veered towards keeping everything jumbled up so as not to attract pests and disease. Having lots of added organic matter in the garden soil will become a buffer against pathogens as soil health keeps improving. My garden of seventeen years has become almost drought proof as when I run out of tank water by October here in Queensland, the plants just sit there for weeks without water. As soon as the rains come the plants perk up again only to continue growing again. Permaculture practices are not hard and fast but if you have thought things through and see the sense of what you are doing in the garden, then the results will come to you.

These days I have very little time to garden but there are always yields enjoy. Gardening kept simple but effective is the key to having success. You can make it as complex or as simple as you like, but most of all it ought to become a part of your daily life. Your vegetable gardens are a natural extension of yourself, after all you are what you eat.

Elisabeth Fekonia

Table of contents

PERMACULTURE Permanent Agriculture

Bill Mollison is the founder of the Permaculture movement with the help of his student, David Holmgren. Together they coined the word Permaculture and evolved a whole new design concept based on traditional agriculture. The objective was to ensure a sustainable way of growing food with multiple yields in a system that will eventually support itself. This was in the mid- seventies, and since then Permaculture has grown to become a global concept that sees solutions within the greater environmental and social crisis that we experience today.

There are the three ethics that Permaculture identifies with

1. Care of Earth
2. Care of People
3. Fair Share

(Mollison page 2 Permaculture Design Manual)

Principles in Permaculture

1. Work with nature not against it
2. The problem is the solution, see the solution not the problem
3. Make the least change for the greatest possible effect
4. The yield of a system is theoretically unlimited
5. Everything gardens
6. Everything works both ways
7. Each element (component) performs many functions
   Make things pay
   Work where it counts
   Use everything to its highest capacity
8. Use relative location
9. Each important function is supported by many elements
10. Design to accelerate succession and evolution
11. Bring food production back to the cities
12. Help make people self- reliant
13. Increase diversity and so create stability
14. Co-operation not competition

(Mollison page 15 Permaculture Design Manual)

Permaculture is a philosophy of working with nature rather than against it. Primarily we need to get our house, garden, our place of living in order so that they can support us. If we do not get our cities, homes, and gardens to feed and shelter us, we must lay waste to all other natural systems. Thus, truly responsible conservationists will have gardens that support their food needs.

(Mollison page 7 Permaculture Design Manual)

With our gardens we can live in harmony with the natural world around us. Permaculture as a design system contains nothing new. It merely arranges that what was always there, in a different way, so that it works to conserve energy or to create more energy than it consumes.

(Mollison 1988, page 9 Permaculture Design Manual)

EVERY DAY THERE ARE MORE REASONS WHY WE SHOULD GROW AT LEAST SOME OF OUR OWN FOOD IN OUR BACKYARD

The world we live in today has never offered us so much in the way of food choices. Food is also relatively cheap and abundant. Why bother to grow your own?

Crops grown with chemicals are not of the same nutritional quality as food was a few generations ago. Somehow the taste isn't quite there either. Since the green revolution began in the fifties, chemical usage in farming has seen a decline in soil health and fertility and human health. Soils that are constantly farmed this way become depleted of its minerals and microbial life. Superphosphate, which is constantly used on our phosphorous and calcium poor soils here in Australia, has acidified the soil and has killed off soil life. Farming with chemicals is working against nature.

Whenever man thinks he has found the answer to overcoming nature, he will find it backfires on him. The irradiation and genetically modifying of our food are two of man's clever strategies to beat the natural odds. Keeping food 'fresher' for longer and to make it grow just how you want.

The only way we can hope to experience good health and vitality is to live close to natures' laws. When we eat naturally grown crops in healthy living soils with our own physical labour, we can truly have a lifestyle that gives us much satisfaction.

There is a strong trend for farmers to walk off the land as drought and subsidised overseas prices become too much of a burden to the Aussie farmer. Before we realise it, we will be importing most of our fresh food from overseas and this means we will be even more reliant on oil and those that control it.

Produce from overseas also has the possibility of being loaded with pesticide residue where regulations are lax or non- existent, and foods that have been genetically modified and irradiated to extend shelf life, are already on our supermarket shelves.

Growing food using Permaculture Principles

There are various schools of thought of how to go about growing your own organic produce. Permaculture principles take a wholistic view that starts from design, to adopting multi-function systems and strategies, and continuous harvesting by intercropping and planting support species. All this terminology might seem a bit daunting to the beginner, but as we go step by step through the planning and design stages it will become clearer as we put the infrastructure into place.

Designing the garden

To be successful at implementing any design, we must look at the most basic elements and work out from there. If you have a block of land to begin with this would be the ideal, as the siting of the house, extra buildings etc. could be incorporated into a fully streamlined and optimal system.

A well-designed property is a marriage of landscape, people, and their needs. Try to put a system together where everything helps each other as much as possible.

Take individual components and assess all their possible functions. Remember the principle of everything working at least both ways?

Example: Chickens give eggs and meat, and they also clean up pests and weed seeds.

General property design

Consider the elements on the land

- ◆ Access and any other earthworks
- ◆ Housing and buildings
- ◆ Water supplies
- ◆ Energy systems
- ◆ Forest, crop, and animal system placements

Sectors

Wild energies that come from outside our system and pass through such as sun, wind, rain, fire, and water flow will either make or break a good design.

On your garden design note the following factors:

- ◆ Fire danger sectors
- ◆ Cold winds in winter
- ◆ Hot, salty, or damaging winds
- ◆ Screening of unwanted views
- ◆ Winter and summer angles
- ◆ Water runoff through the land
- ◆ Flood prone and boggy areas
- ◆ Reflection from ponds and any other water bodies

Elements and their functions

Choose any of the following and use them to design the gardens you wish to create. Draw a mud map of the land surrounding the house and place these elements where they can serve as many functions as possible.

- ◆ Dam
- ◆ House
- ◆ Swales
- ◆ Shed(s)
- ◆ Water tanks
- ◆ Vegetable gardens
- ◆ Orchard
- ◆ Nursery
- ◆ Food forest
- ◆ Compost heaps and worm farms
- ◆ Herb garden/ spiral
- ◆ Banana circles
- ◆ Frog ponds
- ◆ Chicken pen and/or tractor
- ◆ Livestock forage and mulch

Design around the House

The house needs to be sited according to the aspect that is most suited to the property. Considering prevailing winds, water flow, sun/ shade etc. Planting strategically around the house when there are problems with hot summer sun or unwanted views can be remedied with a wind break, or a creating micro- climate. A lot of problems can be solved with good planning and design to cover a multitude of sins.

Plan first then plant as this will make the house a more comfortable place to live in. Deciduous trees can create shade to screen out the hot summer sun and allows the sunlight to pour in with the change of seasons. But leaf litter will either provide good mulch for gardens or become a menace in the gutters. Anticipate both pros and cons with every decision made. You would need to consider where you would place a herb and vegetable garden using the kitchen door as the starting point. When these are planted close to the kitchen, it will be so much more user friendly for the cook. You will also need to look for a handy place for the worm farm or/and the compost heap as well. Or do all your food scraps go to the chickens? If so, then the chicken pen needs to be close enough to the house but far away enough not to be a nuisance.

How about all those spaces between the garden beds? Why not grow a ground cover instead of lawn? This will avoid mowing all those little areas of grass in between gardens that makes mowing or whipper snipping such a chore. There are quite a few useful and hardy ground covers to choose from, and a favourite Permaculture ground cover around vegetable gardens is pinto peanut. This legume plant is very suitable to grow on garden pathways. This helps to increase soil fertility around the garden soil. Pinto peanut will become a carpet like cover. It needs semi- shaded areas and adequate moisture. The pinto peanut is more suitable to plant around raised garden beds such as tank gardens so as not to encroach into the garden beds.

Some of the functions of these elements are as follows:

A dam can supply you with water for the gardens and a breeze from the right direction will cool the air around it. Can this be utilised to give you cooler breezes on hot summer days? Water attracts life and this would encourage frogs to populate the gardens areas where they will find an abundant supply of food, ie. insect pests. A larger dam can also create glare; does it need to be screened out from your favourite viewpoint?

An orchard can give you an abundance of fruit but fallen fruit can create fruit fly problems. Could you accommodate some chickens (or pigs) in the orchard to keep the fruit fly in check and raise meat and eggs at the same time? What about the free fertiliser that comes with this package deal? Growing citrus trees can give some problems with caterpillars destroying the trees. Plant crotalarias in between the trees as these will act as decoys and save your citrus from being demolished. Crotalarias are good legume support plants, and they can be regularly chopped and dropped around the fruit trees to suppress weed growth. This will create a mulch layer to conserve moisture as well as add nutrients for the trees to feed on.

'Chop and Drop'

A weed barrier planted around the gardens will give you a constant supply of mulch material. Something I learnt early on was to grow your mulch where you need it. Such a simple word of advice that makes so much sense. If your mulch material needs to be imported constantly it will most likely cost dollars and effort to get it to where you want. A weed-barrier will not only keep invasive grasses and weeds at bay but will also give you a constant supply of high nutrient mulch.

The classic Permaculture weed barrier is as follows:

An outer row of pigeon pea- Cajanus cajan
A row of arrowroot- Canna edulis
A row of lemon grass- Cymbopogon citratus
A row of comfrey- Symphytum officinale

All these plants will give you a lot of 'chop and drop' biomass in the growing season when growth is at top speed. Cut them back periodically and throw the mulch down onto the gardens. Permaculture gardening in the tropics and sub-tropics utilises plants that have rampant growth for this very purpose. No ground should be left bare in a sub-tropical and tropical climate. Heavy rains will otherwise leach and wash the soil away. Not only does the weed barrier have these two main functions but pigeon pea, arrowroot and comfrey are also excellent forage food for all livestock.

Pigeon pea is one of many legumes that can fertilise the soil by fixing nitrogen from the air, but this is not always the case. Pull up a seedling to see if there are small nodules growing on the roots. If you don't see any then it means the right bacteria for this legume is not present in the soil. An inoculant with the correct rhizobia to develop the nodules on the roots can be bought from a produce store or a plant nursery. These come in sealed, refrigerated packets and can be watered into the soil at around sunset so the hot sun won't kill the bacteria before they make their way into the soil.

The peas that form on the pigeon pea can be eaten and is classed as dahl or split toor tuvar dahl. Harvesting pigeon pea is easy and takes no time at all. Wait until the pods have dried on the shrub then pick them and place them into a pillowcase. Grab the opening of the pillowcase with your hand, close it and trample the contents with your feet. All the broken pods will sit on the top while the peas tend to fall to the bottom corner of the pillowcase. Separate them after a bit of a shake and flick off the dry pods. Finish clearing any remaining debris by winnowing (blowing gently) the chaff into the air. The peas then need to be put through a flour grinder set on a very loose setting. This is to split the peas so that the skins come off. These can then be winnowed away. The peas only take a half an hour or so to cook and tastes great.

Arrowroot is also a source of food for us to eat as this corm has a high starch content. The corms need to be peeled first and then boiled until soft. I usually make the cooked arrowroot corms into a cake by adding cottage cheese, palm sugar, eggs, and vanilla, whiz it through a food processor, pour into a greased baking dish and bake until firm. If this doesn't take your fancy you can throw the whole plant, corms and all to the pigs and they will demolish the whole lot! The arrowroot will need to be 12 to 18 months old to have the starch matured into the corms. Any younger and the corms will taste watery. When the arrowroot is too old, the starches will have turned to fibre and the corms will not soften with cooking.

Lemon grass is a favourite with many Permaculture gardeners as it makes such an attractive garden border. It is easy to mow along the edge of a lemon grass border as there are no hard barriers to watch out for.

Make sure to plant it far away enough from the edge of the garden so it doesn't encroach onto the available garden space. But if it does then there is a good reason to keep cutting it back and throw the clippings onto the garden bed for mulch.

Cutting the lemon grass periodically for mulch is perfect as it is not bulky like pigeon pea and arrowroot. Lemon grass used a mulch is like using regular hay mulch except it has more nutrient value and it is right there where you need it. Do not forget to cut some for a well-deserved cup of tea when you take a gardening break.

Comfrey, the inner row of the weed barrier is an excellent source of mulch as it gives a lot of nutrients to the soil. The comfrey roots grow deep into the sub-soil to mine its minerals. When picking the leaves and throwing them onto the garden you know that you are giving your annual plants the sub-soil minerals, they can't reach for. Serious permaculture gardeners can never have too many comfrey plants growing around the gardens. Comfrey is also an excellent source of greens either raw in salads or cooked. Comfrey root is not suitable for internal use, but it can be boiled up in pig fat to make an oinkment or a salve for bruised skin. All livestock will eat comfrey leaves, but some will need to be encouraged to do so at first. Comfrey is also used as a compost accelerator and as a compost tea.

Comfrey compost tea

Take a large bucket and fill with comfrey leaves. Fill with water and leave for a week or two and use the tea to pour over plants. Under no circumstances stir the mixtures with your hand as the smell is something awful, believe me.

The nursery is an important workstation that needs to be linked up with access from the car, gardens, compost or/and worm-farm and water source. Will there be enough sun? Think of anything else that you will need in the nursery that is required for plant propagation. Visualise yourself working in the nursery, what will you need to have access to? Can you easily transport the seedlings or trees to where they are to be planted? Remember always to calculate where you need to wheelbarrow from point A to point B. One simple rule we have learnt on our steep block is never push a full wheelbarrow uphill.

Gardening with Chickens

The chook run, as mentioned before, is best placed uphill from the garden to catch the nutrient run off from heavy rains. If you plan to use a deep litter system, can you grow enough forage and mulch for the chickens? Remember the food you can grow for the chickens will not only save you money but will give the girls optimum health. Chickens need lots of greens and legumes such as pigeon pea. This will give them more protein so they will keep laying more eggs for you.

Several runs from the chicken housing can be constructed, and they can have access to them one at a time. The other runs remain closed as you grow crops. The main vegetable garden can be constructed this way. This means minimum work and cartage all round.

Chickens do what they do best, and that is to scratch. A chook that cannot scratch is not a happy chook. All that scratching power is often an untapped source of labour, and you might as well make the best of it.

A deep litter system is best done in a fixed run as a deep layer can be built up over time simply by throwing in lots of 'chop and drop' material. This can include garden weeds, old plants, shredded paper the odd bale of hay and cow manure.

A rich mulch layer will eventually be the result, and this can be put onto the garden or compost heap. There are two schools of thought on this as some say all the litter should remain dry so it will turn into a fine powdery mass and keep the chooks from getting parasites, whilst others think that out in the open where the rain can get to it will do the job just as well. The bonus of an outside deep litter system is that worms and insects are found underneath the mulch layer because they are attracted to the moisture of the decomposing material. This is of course a great source of protein for the chickens, hence cutting down on their food bill.

Chook Dome or Tractor

Chickens can also be used to cultivate your gardens prior to planting. A chook dome is an ingenious way of having control over chickens in the garden as they give the benefits of scratching and cleaning out weeds without the headaches and destruction they will cause if they were to free range.

A dome is made of poly pipe and chicken wire. The size of the dome will determine the size of the circular garden shapes.

The instructions for this can be found in Linda Woodrow's book, "The Permaculture Home Garden". The principle of this system is that a series of seven circles evolve to become vegetable gardens. Start with the centre circle and then after several weeks when the chickens have cleared the vegetation, the dome is then moved to another position and so on. You can sheet mulch the area prior to planting out and end up with a minimum fuss vegetable garden. The dome is then revolved around in circles whenever it is time to move to the next garden bed. You will then have maximum weed control, a built-in fertilising system and happy healthy chickens. Whichever way you approach the situation, it is up to your imagination to make the girls work for you. A word of caution though, in our hot summers it maybe too hot to have the chickens in a chook dome when there's only a tarp to cover them.

Growing livestock forage is an important resource as this is the cheapest way to supplement feed for any livestock. It is also the healthiest feed for them as it is fresh and green for optimum nutrition. Grow the forage in the weed barriers, as ground covers such as sweet potato or as support species.

Mulch that consists of grassy types of plants are very suitable as by their very nature they let in the air and moisture. Sugar cane, cow cane, vetiver grass and lemon grass are the firm favourites in Permaculture systems for mulch.

Swales are an ingenious way of keeping water on your land for as long as possible. By digging shallow trenches with a bund (raised mound of soil) on the downside of the slope, the runoff rains will be temporarily paused into the trench and then absorbed into the deeper earth on the mound. The trenches need to be dug on contour for this to be successful. A laser level or an A frame can be used to determine the contour, and the points are then marked out with stakes or white flour or lime for guidelines. A bulldozer can then be employed to do the job of digging a shallow trench about half a metre deep and one metre wide. This can be a major earthwork project and needs to be planned before any infrastructure is put into place.

Banana Circles are designed to water and feed banana plants without the effort of physically looking after them. Dig a hole about two metres across and one metre deep. Place the dirt from the hole around the perimeter to create a mound. Plant the banana suckers/ corms on top of the mound and start to fill in the hole with 'chop and drop' mulch. As the bananas grow it is important to keep them from clumping too much together. A mother, teenager and baby are quite adequate, as they will each in turn have a chance to grow to maturity without robbing nutrients from each other. When a bunch of bananas is harvested, the entire tree can then be chopped down and disposed of into the hole and this will in turn break down to provide its own compost. I have seen an outdoor shower made inside a banana circle and it worked very well.

Herb Spirals are a novel way of growing herbs with different growing requirements all in one cubic metre. The Mediterranean herbs need a well-drained position and mint needs a damp spot to grow in. All the other herbs fall somewhere in between.

The spiral garden is sited to face north and the bottom of the spiral ends in a small pond on the southern end. Start from the ground on a one metre square base and begin on the outside edge from the pond upwards and work your way up into a spiral shape towards the centre to create a height of one metre. Larger rocks are used to build the foundation and form an inside and an outside row. As it is built up, the soil is filled inside the two rows. Stack the rows of rocks one on top of the other in a double row whilst gradually building up the height and filling it with soil as it gets built upwards. The top of the spiral should end up somewhere in the middle, one metre off the ground.

On completion a rosemary or lavender can be planted on top of the spiral. These woody herbs need good drainage, and it also affords them to put their roots deep into the ground. The rocks within the herb spiral also keeps the soil warm and the soil depth gives excellent drainage. Sun loving herbs like sage and thyme can then be planted down from the top of the spiral facing the north and it will be best to finish with the mints and other herbs that need more moisture, behind the rosemary/ lavender which is on the south side. Herbs that require different growing conditions will be happy when placed strategically on the herb spiral.

Herbs suitable for a herb spiral:

Sunny Dry
Shady Dry
Sunny Moist
Shady Moist
Sunny Wet
Shady Wet
Aquatic

Examples of these are, starting at the top of the spiral, Mediterranean herbs such as rosemary or lavender then oregano, sage, thyme, tarragon, coriander, parsley, chives. chamomile, mint and then water chestnuts or arrowhead to grow in the little pond at the bottom of the spiral.

A small pond can be made by using a washing up bowl or alternatively a tyre pond can be constructed and placed at the bottom of the spiral garden. The idea is that any run- off from the spiral garden will end up in the little pond. If there is a concern for mosquitos breeding in the water, duckweed can be placed into the pond where it will form a dense covering over the water surface.

Constructing a tyre pond

Using an old tyre to make a frog pond is a very creative way of recycling such an 'unwantable'. An old truck or tractor tyre that is not steel- belted, will make quite a sizeable pond but any size tyre will do. Dig a hole that is slightly larger than the tyre and taper it down to about two feet, then line it with thick wet newspaper and over the paper line it with some swimming pool liner or builders' plastic. The tyre needs to have one of its sides cut back with a heavy knife. Water thrown over the tyre will aid lubrication to the cutting process, and this will make it easier to do the job. Place the prepared tyre into the hole on top of the liner with the cut off edge facing up and bring the ends of the plastic over the top edge of the hole and trim of neatly. Place rocks or bricks in a circular form around the pond to cover the plastic and help keep it into place. To finish with plant around the edge of the rocks for the frogs to hide in.

Overflow of Water tanks are useful wet areas to grow mint and other moisture loving plants. The overflow can be utilised to grow bananas and banana circles can be planted down from the overflow. This will help to soak up all that excess water thus preventing a nuisance wet area at the same time.

You can also grow some soapwort around the garden tap as this can be handy for washing your hands after doing some gardening work.

You now have some design tools for your house and garden. These systems and infrastructures should be a starting point for you to make gardening a productive and pleasant experience. Remember, you are the manager and only you will know what your needs are. No two gardens are alike. The requirements of any given garden are unique, learn to observe what nature is trying to tell you. Grow what is in season and it will become easy to have healthy plants in abundance.

Different types of gardens

- Raised garden bed
- Mandala garden
- Keyhole garden
- Vertical garden / tipi trellis / pergola
- Herb spiral garden
- No dig-garden
- Chook tractor/ chook run garden
- Tank garden
- Shade cloth garden
- Food forest garden
- Swale garden
- Grey water garden
- Roof top garden
- Water garden- frog pond
- Mulch basket / drum garden
- Pit garden
- Narrow bed
- Broad bed
- Hills and furrows
- Barrier hedge
- Strawbale garden
- Banana circles
- Wicking garden bed

Pit garden- good for sandy soils:

1. Dig a pit such as the banana circle concept, fill with organic waste and allow to break down. Plant when it's broken down like compost.
2. Or line with heavy duty plastic and backfill with soil and organic matter.

Roof top garden

1. Use a vine to grow over a pergola to create a cool shady area on the roof of the house for a cooling effect.
2. Build an attached green/ shade house to help cool the house on the western side.

Mulch basket garden:

1. Dig a hole in the ground then place a wire mesh cylinder on top of the fresh soil and keep filling the basket with compost, soil, old manure, worm castings etc. Grow vines such as tomato plants.
2. Place a spoke wheel on top of the mulch basket garden and grow the vines over it.

Narrow beds:

1. Narrow beds close to the house and herb spiral are easy access to pick the greens.
2. Path side plucking vegetables such as long bearing plants, eggplant, silver beet, celery, kale, mustard greens etc.

Tomato beds could be narrow so they can be reached from both sides. They can also consist of zucchini, peas, eggplant etc.

Broad beds:

These are more suitable for longer maturing plants such as corn, melon, onions, garlic etc.

Wicking garden beds:

Use an old water tank with an open top or construct a water- proof structure at least 60 cm high. Place a PVC downpipe or similar into one corner of the container with an elbow fitted on the end inside the container. A drainage AG pipe can be fitted onto the elbow and laid flat on the bottom. Half fill with gravel, scoria or any other stable material that won't break down in water. Lay a sheet of geo textile fabric over the gravel then shovel the soil media on top. Drill drainage holes level where the soil media meets the gravel, so the soil doesn't become saturated. Make sure the wicking garden is standing completely level. When moisture is needed, place the garden hose into the pipe to top up the water reservoir.

Sheet mulching

Newspapers have been widely used by Permaculture gardeners for decades, and it has proven to be a very effective way to create no-dig gardens. Using newspapers will directly recycle them back into the ground and restore carbon to the soil. Due to changing times in our digital age there is a decline in newspapers, and it has become difficult to find them in volume. Cardboard is also very good, and this resource is still available in abundance. It's a bit more difficult to handle but nonetheless works very well.

What is sheet mulching? Sheet mulching is the no dig method of gardening, and it is very successful if it is done properly. By soaking newspapers in water and laying them on moist ground, a smothering blanket can be created to keep the soil damp and kill off all the weeds and grass underneath. The newspapers/ cardboard will need to be well covered with mulch hay so that it will stay damp to kill the weeds and grass underneath. It is important to generously overlap the wet newspapers and to have them thick enough to create a smothering blanket over the ground. Any gaps will allow the existing weeds to find a toe hold and continue to grow over the newspapers, so it's important that the job is done properly. The following pointers will ensure a successful start to your no-dig garden.

- ◆ Collect newspapers/ cardboard before you aim to start. Ask your neighbours to start saving their old papers for you.
- ◆ Wait until it has rained or water the ground to moisten it. This will help to break down the weeds and soil underneath and it will attract the worms and bugs.
- ◆ Soak the newspapers thoroughly for at least ½ hour before using them. You can use wheelbarrows, wheelie bins, an old bathtub or anything that will hold water to soak the newspapers in. Discard the

glossy inserts as these won't break down efficiently and will have heavy metals in the coloured inks. Most newspapers use soy-based inks for the black print, so these are safe.

- Use them in half newspaper thickness for larger areas (I just open them up) so that the weeds will be kept at bay for many months.

- Use a generous amount of mulch hay in 'biscuit' form not fluffy mulch. You will see that the layers in a bale of mulch hay will come off in sections, hence the name biscuits. Make sure there is absolutely no newspaper visible because if you can see it so does the sun. Any exposed newspaper and the sun will suck up moisture like a wick and you can end up with paper mache underneath the mulch.

- You can scatter blood and bone, dynamic lifter, or manure on the ground before laying down the newspapers. This will encourage the worms and microbes to break down the roots of weeds and grass and break open the hard soil underneath the sheet mulched area. You can also lay down lucerne mulch on top of the paper and then cover with mulch hay.

- Wait at least 3 weeks before planting as by then the weeds would have been smothered and died off. This is to ensure that when you plant and break open the newspaper underneath, you won't have weeds growing through. In the cooler time of the year, you will need to wait a bit longer.

- It's possible to plant straight away into the newly laid sheet mulched area and you can do so by making a hollow in the mulch, use compost to fill the hollow and then plant into the compost. The newspaper needs to stay intact otherwise the weeds will eventually takeover. The disadvantage with this method is that the plants will need regular watering as they dry out very quickly. If the newspaper is laid very thick it will also take far too long for the roots to be able to work its way into the soil, so this needs to be kept into consideration.

Now you have the foundation for the garden, and you are ready for the fun part. Design the pathways and garden areas. You can use your imagination as to how to meander your pathways throughout your garden, but the lie of your land will probably dictate much of that. Your own knowledge of the land will determine where the deeper pockets of topsoil are that are more conducive to direct planting into the ground.

If your soil is poor, then you may want some built up garden beds so the vegetables will have some soil depth and moisture retention for them to grow in. The preparation or condition of your soil will determine how fast or how slow your plants will grow, and a good foundation will certainly help with productivity. Compost, deep litter from the chook pen and well-rotted manure are all good for the vegetable beds before you plant.

Compost is vital and ought to have a great variety of material to ensure that all the minerals for the plants are included. The compost should be well broken down and the finished compost will be about 40% of its original volume. Good compost contains much in the way of bacteria and fungi.

Compost activators are always a good idea as they encourage the decomposition process. Microbes from the soil will also play a huge role, multiply in great numbers, and help with the breakdown process. Following biodynamic principles is very much in the same vein. The formulas that are used in bio-dynamic farming create these microbes in a concentrated form and are dispersed over a large area. Decaying plant material in the soil will then be food for these microbes where they will grow and multiply. This is the foundation of healthy, living soil.

Food grown with chemical input does not have these checks and balances. Herbicides and fungicides are applied to the soil before planting the seed. This has the unfortunate effect of killing the living organisms in the soil. Fertilisers such as N P K do not bring trace- minerals into the soil and these could be sadly lacking. Plants will still grow and look good, but it is what goes on inside the plant that makes a difference. CSIRO tests have confirmed this with their analyses, organic versus chemically grown food have different readings in nutrition.

The life force in chemically farmed soil is not the same as good organically managed soil and eating food from such mineral deficient soil means that we short-change ourselves. Another noose around the neck of good health is all the refining and processing that goes on with our daily food. This takes its toll on some of the nutrients that might have been there to start with but got lost along the way. Is this a form of abuse? I think so. Do we really think that we are nourished enough with the great variety of food we have? Surely, we must pick up what we need along the way?

Not so according to many integrative health practitioners and soil scientists. Most people around the globe are mineral deficient these days.

Enzymes need zinc to do the job of digesting and breaking down the food we eat. If this mineral is missing, then our digestion will be compromised. If we don't digest our food properly, we don't get all the nourishment from our food. If we don't have proper nourishment we don't function as well as we should. For several generations we have been eating refined and denatured food grown in mineral deficient soils with the addition of herbicides, pesticides, and artificial fertilisers.

Our public health system is groaning yet seeks to rectify our short- comings with drugs. We are abusing our health and well-being by eating sub-standard food, and there's no doubt about it, there's always a price to pay. Statistics tell the story.

- One in four will have heart disease,
- one in three will have cancer
- and one in five will have diabetes some time in their life.

Are you motivated enough to take your health and well-being into your own hands? Let us go back now and have a closer look at making good compost so we can succeed in growing our own healthy organic produce.

Making Compost

There is nothing new about making compost. Composting is a practice that has been going on for thousands of years, and nature does it all by itself. Yet many people seem to be a bit afraid of not getting it right and hesitate to have a go.

What is compost?

Compost is a mixture of rich organic matter that has been allowed to rot down. The best example of composting is to look at nature at work on the forest floor. There is a natural yearly cycle of growth and decay. Insects, worms, fungi, and bacteria are constantly at work to create humus. This is what makes a living soil. The soil that has a constant activity of growth and decay, is self-sustaining in cycles of rapid and slow decay throughout the changing seasons. Making your own compost is mimicking nature. You are simply speeding up the process of decay by creating ideal conditions for decomposition to occur.

What makes good compost?

Compost can be composed of any organic material that is piled up high enough to generate some heat, has access to oxygen and keeps enough moisture within itself to encourage microbial activity.

Good compost will smell sweet as the earth and will look like a dark crumbly mass. There should be enough moisture without being too wet.

A combination of thirty parts of carbon to one part of nitrogen will make up a good composition and this means roughly half-green and half-brown parts of dry matter. All plant life contains both elements, but one may be more dominant. Compost becomes colloidal as the slimy trails from worms and snails that feed on it keeps the nutrients together and avoids leaching in heavy rain.

The Science of Composting

A compost heap contains a wide diversity of living matter. Soil organisms are the main catalysts in the heap breaking down compounds into carbohydrates and proteins through enzymic digestion. The carbohydrates break down into simple sugars, organic acids, and carbon dioxide. Proteins are converted into peptides, amino acids, ammonium compounds, atmospheric nitrogen and finally nitrate, which is a form of nitrogen that is available to plants. Bacteria need sufficient oxygen and moisture to survive and when conditions are favourable in the compost heap, they can multiply at an amazing rate by reproducing themselves every five minutes.

Bacteria need carbon for energy and nitrogen to grow and reproduce. They get the energy from oxidising the carbon (turning it into carbon dioxide) and the heat in the compost heap is the result of this oxidation as the bacteria burn up the carbon. Getting the right balance of carbon and nitrogen is therefore crucial whereas if the bacteria have too much or too little of either element they will die. Think brown and green when trying to identify the two elements for the compost heap.

Carbon rich sources:

Straw, leaves, woody matter, corn stalks etc. are brown

Nitrogen rich sources:

Fresh grass clippings, kitchen waste, animal manure etc are green

30 parts of carbon to 1 part nitrogen

All plants contain both elements, but one may be dominant.

- Eg dead leaves 40:1
- Fresh grass clippings 20:1

These two mixed together will create 30:1 ratio, which is ideal. Air and water- keeps the heap aerobic. Turn the heap at intervals.

Let's build a compost heap

If you have a selection of material at hand to build up a new pile, place the coarsest material on the bottom. Any cornstalks or twiggy branches should go down first then a layer of green matter such as grass clippings or green 'chop and drop'. A thin layer of manure followed by some more coarse or dry matter and so on.

Compost has a great food supply for bacteria, but adequate food for the fungi is often lacking. Having fungi present in a compost heap is imperative for a healthy balance of soil life. Fungi needs carbon rich material to feed on and multiply. Adding woody components throughout the heap is good but these should be shredded first to ensure maximum surface area for the fungi to feed on. Both fungi and soil bacteria help plants to grow by supplying nutrients for healthy growth.

It is crucial to wet down each layer thoroughly or soak the material in water prior to assembling the heap. You can add some diluted molasses to feed the microbes so they can quickly increase in numbers.

Another way of compiling the ingredients for the compost pile would be a bucket of kitchen scraps; half a bucket of manure mixed with hay, two wheelbarrows of weeds and old plants from the garden and half a wheelbarrow of fresh, green grass clippings.

Generally, it is two parts of plant waste to one part manure, one part green to one part brown. Seaweed, comfrey, yarrow, nettles, and urine are good compost activators as well as road or home kills of any small animal.

A sprinkle of wood ash is fine but be careful, do not add too much. Leave out any chemically treated wood.

The size of the compost heap is also important for fast decomposition, as certain micro-organisms do their work at various temperatures. A minimum of one cubic metre is required and it can be a free-standing heap or built into a box with three sides. If the compost is built up into a compost bin or box, it is easier to keep the sides of the heap moist ensuring even decomposition. A cover of some sort will be a good idea as nutrient leaching will occur with heavy rain. These are the main requirements to make a good compost heap.

The next step would be to turn the heap at regular intervals. Turning the heap will add oxygen and mix the contents throughout to encourage even decomposition. If you are like me and never seem to get around to turning the compost heap, do not despair, it will become a passive compost heap, and will take longer for decomposition to occur.

Summary of ingredients for the compost pile

- Anything that was once living matter can be put into the compost heap
- Two parts plant waste to one part manure
- One part green to one part brown
- Soft garden prunings, seaweed, comfrey, yarrow, and nettles
- Newspaper, paper, and cardboard (shredded)
- Dry leaves, straw and old plant remains (chopped)
- Wood chips, bark and saw dust are very low in nitrogen and take a long time to break down
- Wood ash, a little bit sprinkled at intervals is fine

Summary of suitable ratio of carbon-nitrogen

- 1 bucket moist kitchen waste
- ½ bucket old horse manure mixed with straw
- 2 wheelbarrows with old plants from the garden, shredded
- 1 bag of hedge clippings, shredded

Materials to avoid in the compost

- Fats and oils
- Weeds with seed unless it becomes a thermophilic compost
- Diseased plants
- Chemically treated wood products
- Coal or charcoal ash
- Dog and cat faeces unless it is a separate compost

How decomposition takes place

Almost immediately the action starts. Psychrophiles start working at 6 C. These start to digest carbon compounds and will alter the chemical state of organic matter giving of small amounts of energy that will eventually raise the temperature of the pile for the next group of bacteria to thrive, the mesophiles, a mid-range temperature bacterium that thrive at 30-39 C. These do most of the work of decomposing the heap. Even if the heap does not reach higher temperatures than this, it does not really matter as most decomposition takes place at this temperature. Larger creatures such as worms and insects are working in tandem with the mesophiles breaking down organic matter into smaller pieces by eating and digesting it. It becomes easier if matter has been shredded beforehand where more surface area is exposed to the bacteria and decomposition will speed up.

If conditions are right, enough air, water, carbon/ nitrogen balance and the pile is big enough, the temperature will rise even further to 44- 70 C by the thermophiles.

Decomposition is now in full swing, and the heat can get very intense. The thermophiles can maintain this temperature for 3- 5 days before dying back. Turning the compost heap at this stage will provide a new fix of oxygen and a new cycle can begin again, up to three or four times.

The advantage of such a high temperature is that disease and weed seeds are destroyed. Once the heat drops back down the microbial activity starts to decrease and other organisms take over to complete the breaking down process. At these mid temperatures the fungi such as actinomycetes and streptomycetes are at work producing natural antibiotics that keep disease at bay. You know these organisms are present when you see white cobwebby structures, a sign of good healthy compost. In perfect conditions the compost can be ready in 6 weeks, but it will more likely be several months.

If compost is used very fresh it can still have nitrogen draw down as it is still decomposing instead of being able to give nutrients to the plants.

When is the compost ready to use?

Take some compost in your hand and squeeze it and drop in it on the ground. If your hand is clean after handling the compost that is a good sign. It should smell fresh and earthy and have a black colour.

If you see streaks of matter on your hand, then it's not quite there yet. The pH of compost should be neutral or pH 7.

The cool heap

Passive. Pile organic matter in a heap as material becomes available. Six months to two years for finished compost.

Hot heap

Active. Manage the heap carefully to achieve heat and decomposition. Can be ready to use in a matter of weeks.

Compost Activators

If there is no manure in the compost heap it may pay to add some activators. Urine, comfrey and yarrow, a roadkill or even some leftover meat, fish remains, are all good activators for the compost as they are high in nitrogen. A spade full mature compost also helps. Make sure to dig any meat, fish, or small amounts of dairy well inside the heap to avoid rats invading the compost.

Common problems

If the heap starts to smell unpleasant there may be too much nitrogen rich material. Add some dry straw, dry leaves etc. or if it's becoming anaerobic, there isn't enough oxygen, turn the pile.

If the compost heap smells sour, it might be too wet. Make sure it is covered in heavy rain.

The compost heap is too dry- nothing seems to be rotting down-does it need moisture? Add some compost activators such as diluted urine. Next time you turn the heap you might like to place a dead chicken etc. in the middle towards the bottom of the pile and add some more water if necessary.

Here's a good tip to see when compost is ready to use. If toads are attracted to your compost heap you can be sure you're doing it right. Toads are attracted to the bugs that are flocking to the compost once active decomposition has slowed down, so next time you see them around do not curse them, they might be trying to tell you something.

Worm farming

Manure (and humanure) decomposed through a worm farm is another excellent way to fertilise the soil. If you don't have easy access to manure, you'd do well to have a compost toilet. Humanure is a great asset that should not be left outside the cycle of sustainable gardening. How wonderful it would be if everyone would recycle all their human waste. Waste is not really a waste but rather a valuable resource. We ought to actively recycle all organic waste through decomposition and use it in the garden to grow our food. This is truly sustainable agriculture. We would build up the fertility in our gardens if everyone cycled all their organic waste.

Unfortunately, the possibility of this happening is still rather remote as most people have septic or sewerage systems. That is something that must change as well. Gardening successfully without using manure does become difficult because manure carries the life force that constantly keeps on introducing huge populations of microbes into the soil. The microbes, fungi, yeasts, enzymes, and a myriad of other organisms are the digestive system of the soil. These continuously release food for the plants through the breaking down of organic matter. This is how we arrive at a chain of events that brings life and death constantly throughout the soil.

Fertility and decay are two sides of the same coin.

Worm castings and compost are colloidal and that means they can hold up to 10 times their own weight in water, they have a high CEC rating (cation exchange capacity), and they help to create good soil structure.

Compost Worms

It has been estimated that one teaspoon full of vermicast has up to four billion bacteria whose life cycles interweaves with countless protozoa, fungi, worms, and many other creatures. These all form the web of soil life that is the basis of plant vitality. There are 5 – 10 times more available plant nutrients in vermicast than there is in compost or soil.

- ◆ Vermicast is deemed to be pathogen free
- ◆ Composting with worms takes less time than making compost

A variety of worms used for compost worm farms are the tiger worm (eisenia fetida), red worm (lumbicus rubellus), the red tiger worm (eisenia andrei) and the African night crawler.

The reason for this is that the common earth worms are not easily held in captivity, and they tend to go deeper down into the soil. Compost worms are by nature surface dwellers. They do need particular attention as they won't survive in an open garden due to our hot climate. Compost worms are not native to Australia so do not expect to have them thrive for any length of time outside of the ideal conditions of a worm farm.

Compost worms need a confined container that will retain moisture. The worm farm needs to be sited in such a way that is out of direct sunlight and in a shaded position, preferably with a southerly breeze around it. Moisture needs to be kept up to them and they need a constant supply of food.

Setting up a worm bed.

There are varied commercial worm farms available, and they all do the job. It isn't necessary however to buy a worm farm as it is so easy to make your own.

For the beginner you will do well to find a styrene broccoli box. Drill some drainage holes in the bottom of the box where the bottom meets the front. Place some fly screen material or shade cloth over the drainage holes to prevent the worms from escaping and tape into place with some duct tape. The lid needs a sizable square hole cut into it for added ventilation and this also needs to be covered with fly screen mesh and taped into place. You can now make the bedding ready so the worms will be comfortable in their new home.

- The bottom of the worm bed should have a few sheets of wet newspaper lined on the bottom.
- Crumple up some more newspaper and mix in some soil that is high in organic matter or matured vermicast or compost and add manure or/ and food scraps on top of this.
- Make sure that the styrene box is slightly tilted forward for drainage, and you have a bucket placed underneath it to catch the liquids. These are your liquid worm castings. Dilute this liquid at a ratio of 1:10 with water and use over your plants, especially seedlings, to encourage healthy plant growth.

Bathtub worm farm

- Fill an old bathtub that is slightly tilted towards the plug hole. Place a bucket underneath to catch the worm juice. Fill with fresh manure (not chicken manure), moisten, and add the compost worms. You will need to keep a check on moisture levels now and then as if it becomes too dry there won't be much worm activity
- Always remember that worms are soft, moist, slimy little creatures and they like their food and housing the same way.
- Finish with a cover of plastic and some carpet or underfelt on top. I've discovered that sheep fleece makes an excellent cover for the worm farm as it keeps the moisture in for many months without having to check it. It is also rat proof as the critters won't be able to poke their head under the wet fleece to go in and gobble up all your precious compost worms. I call it the set and forget method.

Worms control their population in relation to the available surface area and food so they can be expanded into making more worms farms. I've seen people with several bathtubs made into worm farms and transform their poor sandy soil into a garden of Eden. Just keep the drainage requirements in mind and place a bucket underneath to catch all that lovely worm juice.

You will soon see that once you start keeping compost worms, they will keep on expanding into more worm farms.

Food suitable for the worm farm:

Worms like their environment moist and not too warm. In fact, they will suffer and even escape if the temperature is over 25 C. Their very nature is soft and slimy and so their food supply should also reflect this. Any dry matter will not be particularly attractive to the worms as they are not like caterpillars that will chew into the leaves. Any food scraps added to the worm farm should be wilted and moist. Any type of manure is ok for the worm farm, apart from fowl. Humanure is perfect for a total recycling of human waste. Once you have such a system in place, you have all your minerals cycling around on your land and this means soil, plant, animal, and human health will eventually meet all their mineral requirements. This is true sustainable gardening.

I introduced the broccoli box worm farm to a lady living in Caloundra on the Sunshine Coast, on a house block. For several years now she's been diligently recycling her humanure through her garden with great success! No one knows it's happening as it's done very discreetly. That's our little secret, ok?

Pathogens and viruses are eliminated by passing through the worm's digestive tract so the vermicast that is left behind is always safe to use for growing food.

Food for the worms

- Kitchen scraps (except onion skins and too much citrus)
- Seaweed
- Garden waste
- Manure

Potting mix

Vermicast is classed as sterile, so it is very suitable for raising seedlings in. The pure vermicast should never be used on its own as it will set like concrete if it is allowed to dry out. A good potting soil mix would be 50% of vermicast with 50% sharp river sand. Make sure it is (sharp) river sand and not beach or builders' sand as these will act as a water repellant instead. I also like to add vermiculite and cocopeat to the mix to make the worm castings go further.

Foliar spray

A foliar spray can be made by using one part of liquid worm castings to five parts water. This should be well aerated for 24 hours. A fish tank pump can be inserted to produce a constant bubbling effect. This is very important as the bacteria can then grow and multiply when the liquid is oxygenated. Add a cup of molasses to five litres of water and this will aid microbial growth. Spray onto seedlings for a healthy growth rate. This foliar spray can be used to combat powdery mildew and black spot in fruit trees.

Brewing Aerobic Compost Tea

Use only fully matured, fresh smelling compost, a pillowcase or pillow protector, a twenty-litre bucket and a set of air stones and a cup of organic molasses. Some other suggested ingredients to add are seasol or natrakelp, instant humus, rock dust and liquid potash just to name a few.

- Loosely half fill the pillowcase with the compost/ vermicast
- Add clean water to the bucket, chlorinated water needs to sit for several hours
- beforehand to gas off the chlorine
- Turn on the aerator stones once all the ingredients are added to the bucket
- Bubble for one or two days. In 24 hours, the tea should be active and frothy
- Let the brew settle down for 10 -20 minutes and then strain into another bucket
- Use it while it is still fresh

Apply to the soil or as a foliar spray at the end of the day so the sun doesn't kill the microbes before they work their way into the soil

The remains of the compost can be put back onto the compost heap as there are plenty of good bacteria left in them

Using the Tea

How often to foliar spray your plants?

Beneficial insects are a good indicator of your garden's health. If you don't have good levels- spray at least once a month or as often as every two weeks

Start when the plants have developed their first true leaves

To control damping off, spray the soil with full strength tea as soon as you plant

On trees and shrubs- two weeks before bud break, then every 10 – 14 days

You'll need to spray every 10 days if you have a neighbour who sprays pesticides because pesticides kill beneficial organisms as well as pests.

Living soil

The daily miracles that occur in the soil are astounding. Electrical currents, enzymes, bacteria, fungi, and minerals are at the bottom of this wonderful chain of events that gives life to the soil. Positive and negative charges with the aid of moisture for conductivity set of a chain of events that bring the very forces of life into action. The link from rock to living form is brought about by weathering through carbonic acid, the action of catalysts such as enzymes, and through electrical charges.

The foundation of life is in the soil, and we can nurture and sustain it for our benefit. When food is grown in soil that has all the life-giving forces in place, it cannot help but produce healthy plants. The right minerals in the right proportions will be made available to those plants and these will in turn impart their minerals onto us. This is the basis of nutrient dense food. Foods grown with chemical fertilisers have different nutritional readings. They are not as mineral rich as our organic counterparts or as high in sugars. This is due to the way these plants are grown. All plants have two sets of roots. The thin white roots are the feeder-roots, and they take up surface soil nutrients and moisture, and the older darker roots are the roots that take up the moisture and minerals from deep down in the subsoil. Plants, like humans cannot feed directly of the minerals from the soil. Plant nutrients come from the cation/anion interchange to become available. This very complex set of circumstances is provided in the humus part of the soil and all the life forms associated with it. Growing plants without the natural complexity of soil life, but rather by using this water based chemical fertiliser, is to force feed them.

The roots are then forced to take up whatever the farmer applies to the soil and the plant is force- fed with the fertiliser. The overuse of urea to add nitrogen to the soil is also a very real problem to our health. Vegetables that give off a lot of water are a symptomatic of being force fed by too much nitrogen. Large crisp looking cabbages and enormous lettuces that are chemically grown are full of nitrates that form into nitrites and contribute to ill health. Nitrates in our food and water cause diseases such as cancer, methemoglobinemia, and enlargement of the thyroid gland and diabetes mellitus. The entire metabolism of these plants is interfered with as they are giants composed of mainly salts and water. Peeling conventionally grown potatoes is a messy business against the splashboard in my kitchen; it always needs a good wipe down after peeling them. This does not happen when I peel my home- grown potatoes.

A word of caution for organic growers is the practice of directly planting into fresh manure. Raw manure that has not been composted or put through a worm farm or is less than six months old, will make the soil prone to nitrogen overload and the plants will be very green as a result. Have you noticed that cows do not graze near fresh cowpats? This is because the bright green grass around the manure tastes bitter. In time, the bugs and the microbes will break down the manure and the grass will sweeten up again.

I have put raw manure on my garden beds for quite a few years now and not bothered to compost it first. Now there's a contradiction, but I don't plant into it straight away. The manure is put onto the garden beds and allowed to break down before planting. The life in my soil is so active and the humus so well built up that it is teeming with activity. In just a few weeks the manure will have broken down and is ready to be planted without any worry of nitrogen overload. This simply skips the double handling of putting the manure through the composting process first. It is important not to spread the manure too thick onto the garden beds otherwise it's like stuffing yourself with a huge meal and having trouble digesting it all! The soil microbes can suffer from overload as well. If you want to put straight manure onto your garden beds and you are doubtful of the soil life within, you'd do well to speed up the break down of the manure and any other food scraps with some bokashi.

Sprinkle the bokashi on top of the manure and make sure everything is well watered, and then cover with mulch.

The mulch will keep everything moist and prevent gassing off the nutrients, and in just a few weeks the soil should look like soil and nothing else but soil. Don't be surprised if you find lots of earthworms and bugs in the garden.

Bokashi

Bokashi is Japanese for ferment.

It is a compost starter that has been devised by a couple of Japanese scientists. Aerobic and anaerobic microbes are cultured, and these are rather like cheese bacteria. The soil added to the recipe could be your own and some compost or worm castings can be included as well.

All manure and food scraps need to be broken down by enzymes and bacteria and bokashi will gently help speed up this process.

- 1 kg rice or wheat bran
- 1 tablespoon brown sugar
- 1 lb of medium grain white rice
- Several cups of water

Soak the rice in water for a few minutes, swish it around with the hand and drain the rice water into another bowl. Add the sugar or a little molasses to the rice water to assist fermentation. Leave to ferment for a week. It should become active before the next step.

- Mix some of the rice water into the bran and add about 100g of your own soil/ compost and mix well until just moist. The mixture should be just wet enough to stick together when squeezed
- Leave in a plastic bag excluding all air, seal tightly and leave for seven to ten days in a dark place
- Spread onto a tray and allow to dry slowly in the shade
- The bokashi can be stored in a container when thoroughly dried
- Use approximately one teaspoon to a good handful of vegie scraps or sprinkle over the manure on the garden beds before covering with mulch

Homemade Bokashi bin

You will need two buckets, one twenty litre and one ten litre bucket. Drill plenty of holes in the bottom of the ten-litre bucket and insert into the twenty-litre bucket. Fill the inside bucket with kitchen scraps and sprinkle some bokashi over the scraps at intervals. Keep a lid on the bucket until it's full. Take the inside bucket out and empty into the garden bed by digging in the pickled food scraps or empty into the compost. There will be a strong bokashi liquid in the twenty-litre bucket that can be used as a tea.

Tea from the bokashi bin

Bokashi tea is a very nutrient rich fertilizer that can be used on your indoor plants, lawns, veggies, and flowers. Bokashi tea is quite acidic and therefore it's recommended to dilute the tea around 1:100. You may wish to test the dilution rate on sensitive plants, and you may find that less sensitive plants can tolerate a lower dilution rate. The diluted bokashi tea fertilizer should only be applied to the soil as the foliage will be more sensitive to low pH levels. Remember bokashi tea is teeming with the beneficial bacteria. Use your bokashi tea as soon as possible after draining it from your bokashi kitchen composter so that your plants can benefit from all the goodness in it. If left unused for more than a few hours, then the tea may start to go bad…. anaerobic…. and smell awful!

Compost enhancer

Bokashi tea has millions of the microbes from your bokashi kitchen composter. These can be incredibly beneficial to your compost pile and can be poured directly into it. The bacteria will help to speed up the composting process in your compost pile. Again, be sure to use fresh bokashi tea that you have just drained from your kitchen composter, otherwise keep it in the fridge for several weeks with the lid tightly screwed on!

Drain unblocker

You can pour it down the drain as it is completely natural and will not pollute. In fact, the bokashi bacteria and enzymes can help to unblock clogged drains and are beneficial to the water treatment works too.

Basic soil types Understanding your soil

Once you know your soil type, you will understand how water moves through the soil. This doesn't necessarily mean that you scratch the mulch aside and look at the top of the soil for this can be deceptive. Do you know that the roots from the plants are also like a lamp-wick? Moisture is drawn up from underneath the soil as well as from the top. Watering can so easily be over-done, and this does nothing but leach out your precious minerals. You will learn by experience when the garden needs a good watering. Learn to observe and look for the signs.

- Aim to have your soil absorb water but not become waterlogged. Good drainage is very essential
- Compaction can also be a real problem and the soil will need to be aerated

Next time you dig a hole for a tree or a fence post, dig down to about three feet, this will enable you to observe your soil profile

- How deep is your topsoil?
- Are there any differences in colour and texture?
- Mottled grey and rust brown streaks in the lower horizons indicate leached acid soils and possible drainage problem
- Brown and red colours are from oxidised iron, a result of acidity
- Absence of horizons may indicate a very deep topsoil or a good organic content
- Layers of different coloured clays are commonly found in tropical soils
- Looking at your plant profile, how deep do the roots penetrate?
- How far down can you find worm tunnels or other visible creatures?

Note the depth of any compacted layer or hardpan when roots end abruptly. (This is often found in soils that have been commercially farmed, due to ploughing)

The key component of soil vitality is the movement of water, soil life, organic matter, air, and nutrients between the topsoil and subsoil. The nature of the boundary between the two should be uneven, rather gradual with channels made by plant roots well into the subsoil. If you can see that then you have good dirt.

If not here's what you do about it.

Add organic matter!

Just about every problem with soil can be remedied by adding organic matter. Constantly adding compost and mulch to your garden beds is the kindest thing you can do for soil health. Compost keeps on feeding the soil organisms that make a living and biologically active soil. Moisture is retained and the soil temperature is kept stable. Your garden will keep on thriving with all that humus. I add minerals to my gardens via the animal manure. All my milking girls have kelp and other minerals given to them daily. What they don't need will surely end up on the dinner plate via the garden. If you don't have your own source of manure, try worm farming. This kind of livestock is available to everyone. You can also add minerals in the form of 'naturamin' or 'alroc', kelp and calcigrit to the worm farm where all these minerals will end up in the soil. The mineral dust is a fine mixture of a combination of various crushed rocks that can be scattered over the garden beds and in about 6 months time this will be taken up into the soil by the soil life.

- Permaculture is based on the design of traditional cultivated systems modelled on diversity and stability and resilience of established eco systems
- Permaculture is a beautiful blend of traditional knowledge and scientific know how

As a result, there are designs in Permaculture systems that provide feed back loops that can be modified as needed. The best examples of Permaculture can be found in the tropics, in the food forest garden. Living mulches or ground covers are often used in Permaculture systems and these give the ground a shelter from the hot sun, maintain soil moisture and create mulch. Ground covers also help keep weeds at bay. Creating and maintaining humus is most important in a soil management program. It's a constant process. Compacted and poorly drained soil will lack the microbial populations needed to process organic matter. It may just sit there and not break down very much. This is where knowledge on how to make good compost is so important. A well-made compost will inoculate the soil with organisms that will eventually lead to the decay of all organic matter.

Soil life in the form of micro-organisms such as mycorrhizae, moulds, yeasts, and fungi are necessary to start this break down process. This is not unlike our own intestinal tracts where millions of micro-organisms live within our gut. These micro-organisms are needed to break down the food we eat. Decay organisms need a pH range from 6-8. If your soil test comes up with a lower pH that means it is starving of organic matter. Most problems with soil are solved by simply adding humus to the soil. We are encouraged to grow as much mulch as possible in our gardens. Grow your mulch where you need it is the catch cry. Tropical legumes such as pigeon pea, cow pea, lab lab, crotolaria etc. plus all the grass types such as vetiver and lemon grass as well as sugar cane are wonderful sources of mulch material. There is a vast list to choose from in a tropical/ sub- tropical climate. We should be growing plants for mulch all over the place. When a lot of rough mulch is chopped and dropped, manure can also be incorporated into it to speed up the decaying process. Earthworms are encouraged to feast on all that raw material as they will multiply and add their worm castings. Earthworms will aerate and mix the soil and poorly drained and compacted soils can be brought to life. The minimum amount of organic matter in the form of humus should be at least 10%. With all this organic matter, the soil will carry so much life that it simply must improve and become productive.

Soil Structure

The structure of the soil is the way soil particles group together to form small crumbs called peds. In most soils this is very apparent but in soils composed of beach sand, the soil grains do not stick together. Also, the particles of some clay soils that join in a large featureless mass, this is the other end of the scale and both types are said to be soils of no structure. Both sandy and clay soils are remedied by adding lots of organic matter. Organic matter helps to give the soil a good structure because it helps to bind soil particles together and gives spaces in between. Good soil structure is important as this allows water to soak into the soil and allows excess water to drain out. Air needs to be able to move through soil as well. This is very important for the soil life. Soil life needs oxygen to do their job. (Aerobic bacteria and other larger forms of life such as worms and insects). Good soil structure is essential for good drainage and aeration.

Soil permeability is a measure of how easily water moves throughout the soil. It affects the rate at which water can enter the soil, the infiltration rate.

The porosity is measured by how much spaces are around the soil peds that are filled by air or water.

The ideal soil is one that is very productive and will support healthy plants. This can mean any type of soil that has plenty of soil biota and organic matter.

- This type of soil will be well structured
- Is full of nutrient
- Is well-drained
- Is biologically active
- Has medium texture
- Has a pH of 6.3-6.8

A combination of these features will provide an environment that is ideal for plant growth.

Testing Your Soil Texture

A method of testing your soil type is to do a soil test in water.

Place some soil in a jar 4 cm deep and add water to about three quarters full. Close the lid on the jar and give a good shake. Allow the jar to sit undisturbed for at least 24 hours or until the water becomes clear. Depending on your soil type this can take anything from a day to well over a week.

Once the water has cleared hold the jar up to bright sunlight and look for three layers of soil. Out of a total thickness of 4 cm of soil depth, the layers of a good soil will measure something like the following:

2.1 cm sand and grit
1.3 cm silt
.6 cm clay

Result is:

45% sand and grit
33% silt
22% clay

Organic matter can be seen floating on top of the water. If the water remains milky or turbid, this is a suspension of very fine or colloidal clay.

You can do an online search for 'analysing soil texture' to find your soil type.

There are many different soils, but they can be divided into the following groups:

Sandy soil has a gritty feel and does not stick together
Sandy loam is friable and sticks together
Loamy soil is friable and sticks together and sand grains cannot be felt

These three soil types may or may not roll into a ball, but they will not roll into a sausage shape and will break easily

Sandy clay loam has a gritty feel
Clay loam is where the sandy grains cannot be felt

These two soil types will roll into a sausage but cannot be turned into a ring without cracking

Light clay is easy to mould
Medium clay is a bit stiff and is easily moulded.
Heavy clay is very stiff and can still be moulded

These three soil types can be made into a sausage and can also be turned into a ring without a lot of cracking

Clay soils

Clay often gets a bad press but these soils can perform well in the garden, depending on the type of clay and how it's managed. By understanding clay and what gives it its unique properties, it can be turned into a beautiful growing medium. Clay is a special mineral. It is like flat, little, platelets, with an enormous surface area. For example, a gram of bentonite clay has the surface area of a football field. It also has a weak negative charge which is the key to its nutrient and water holding ability. Clay soils are not necessarily bad. With some simple testing and the right sort of treatment, it's possible to turn a nasty, heavy clay into a beautiful crumbly topsoil that will grow virtually anything. The secret is to put every bit of organic matter back into your soil.

Sandy soils

There are two types of sand, beach sand and river sand. One repels water and the other absorbs it. Sand and silt do not have a negative charge, nor do they have the surface area of clay which explains why their water and nutrient holding is so much poorer. If you live close to the beach and wish to garden productively it will pay to get a delivery of sharp/ river sand to pull the moisture into the ground. At one stage I have purchased premium garden soil to improve soil texture, but it was sterile as it had pine bark mixed with chicken manure and it suffered badly from nitrogen drawdown. Find out the ingredients that have may have been mixed in with the topsoil if you choose to buy it. Don't get caught out like I did. A foot note on buying in topsoil. Ask them where it comes from as often new estates are created where they take off the topsoil and the new landowners need to buy their topsoil so they can grow a garden! It pays to make sure your topsoil is ethically sourced.

Cation Exchange Capacity CEC

Soil is made of crushed rock into various particle sizes. It is made of gravel, sand, silt, and clay. These are referred to as cations and they are positively charged except for clay. Organic matter including gasses are negatively charged particles and these are referred to as anions. The exchange of minerals from the soil is the nutrient capacity of the soil. Both clay (the finest particle size of the original rock) and organic matter are colloidal in nature, and both have easily accessible nutrients available to plants and they are negatively charged. It's important to realise the number of positive and negative charges attributed to each cation and anion. There are two storage mechanisms in the soil, the clay colloid, and the humus colloid (colloid is a term for a small particle size). Both clay and humus are negatively charged, which means that positively charged nutrients (cations) can attach to them. Humus also contains positively charged sites, so it also can store anions.

This storage capacity can be viewed in terms of the soil's 'pantry size', as it relates to the storage of plant food.

Nutrient storage capacity of soils
Sand has a poor CEC rating of 2-3
Silt is a little better at 5-7
Heavy clay is good at 30-60
Humus has 250
Humic acid 450
Fulvic acid an astronomical 1400 and is found in well-made compost

Bacteria and enzymes feed on the organic matter in the soil. Enzymes are simple proteins that help breakdown anything that is of organic origin. Some enzymes can use rock minerals straight from the soil to facilitate energy transfer reactions. (CEC) Plant productivity depends on the efficient discharge of these functions.

Soil pH

pH stands for potential hydrogen. If your soil measures a pH of 5.5 this means that there are more hydrogen ions in your soil than a soil measuring 6.5 pH. These hydrogen ions fill the void that nutrients leave behind. Soil that is devoid of organic matter has a low pH and to increase the pH of your soil all you need to do is add organic matter. This will then eventually increase your soil pH. Adding lime to an acidic soil will also help to sweeten it up. If your soil is made of heavy clay, then it is advisable to add gypsum instead of lime as this doesn't influence the soil pH. Clay soils often have a good pH reading and gypsum will help break up the clay over time. Add lime or gypsum a couple of times a year if this is needed. Dolomite is a combination of calcium and magnesium, and both are vital for productive plant growth. Using dolomite could create problems if the ratio of both minerals is out of balance. I have learned that if mud sticks to your boots in rainy weather and making them heavy to walk in, then you have more magnesium in ratio to calcium and therefore it is best to add lime, not dolomite to increase the amount of calcium. When I walk around outside after heavy rains my boots stay clean with only a few streaks of mud clinging to them. This shows me I have a good balance of magnesium ratio to calcium in my soil.

To measure your soil pH a soil test kit is easily available at produce shops and nursery suppliers. An extreme in either direction can spell disaster for garden productivity as it will cause an excess of certain minerals and lock up others.

Soil pH

The effects of pH on soil nutrients

Increasing acidity					Increasing alkalinity		
←						→	

toxic acidity extremely strong	strong	moderate	slight	neutral	slight	toxic alkalinity moderate	strong
3	4	5	6	7	8	9	10

Excess of
   ◆ Aluminium
   ◆ Manganese
   ◆ Iron
	Excess of
	   ◆ Sodium

Deficiency of
   ◆ Magnesium
   ◆ Calcium
   ◆ Potassium
   ◆ Phosphorus
   ◆ Molybdenum
	Deficiency of
	   ◆ Iron
	   ◆ Manganese
	   ◆ Zinc
	   ◆ Copper
	   ◆ Boron

Seed Saving

From the Seed Savers' Handbook
by Michel and Jude Fanton
used by permission www.seedsavers.net

I thoroughly recommend reading the seed savers handbook as it's much more in depth than what is in these pages. Their book is for sale from their website.

Why save our own seed?

Five vegetable seed companies control 67% of the global vegetable seed market. These multinational seed companies develop, breed and market chemically dependant hybrids and genetically modified seeds. This of course is in their best interest. It's hard for a multinational company to make money out of seed that can be grown by organic means. There are no herbicides to market, nor artificial fertilisers to coach the plants to grow lush and abundantly. Open-pollinated seed cannot be patented but genetically modified seed can.

Instead of millions of farmers and gardeners across the world producing billions of locally adapted seed varieties, we have a handful of seed companies selling a handful of varieties. On a global scale, it is dangerous to reduce the diversity of our food base. It will become harder to find pest and disease resistant varieties adaptable to our changing climates. For the home gardener it means having your own locality adapt to a one size fits all. This seed will be grown in countries where labour is cheap, and your own local climatic conditions will need to be like the local conditions those seed were grown in.

For those of us who wish to garden and grow our own food without pesticides and artificial fertilisers, it is often difficult to source seed of crops that have proven themselves in local climates. This situation is only going to grow worse as the seed companies are intent on swallowing up the available open-pollinated and local varieties and replace them with chemically dependant and G.M. seed.

Crossing two genetically different varieties requires a considerable amount of hand labour. A lot of the seed production takes place with minimum wages such as Chile, Taiwan, Kenya, and Indonesia. Broadly speaking, this is the situation where we are heading with our global food supply.

Plant breeders' rights

This is a further threat to the biodiversity of plants through the plant breeders' rights act. It remains to be legally tested whether this legislation could be used to stop seed saving effectively. This means it might become illegal to save and sell locally grown seed!! It's already being enforced in some countries in Africa where it has become illegal to pass on seed more than three villages away. Heavy fines and jail terms will be enforced from 2021 to protect the interests of big seed companies. We must stand up against such bullying as it is a basic human right to save your own seed and sell or swap them with whoever we like. As organic gardeners we must assume responsibility by learning how to save organic, open-pollinated seed to secure healthy food for the present and the future.

What is open-pollinated seed?

In the natural world, flowers are pollinated by wind, insects, and birds. This is the natural way of the reproduction of plant species. When man started doing selective breeding, he found he could manipulate certain characteristics that he favoured and hence we have the food that we eat today. By selectively breeding through hand pollinating, he could breed plants for colour, taste and texture. In the wild, certain plants crossed because they were closely related, hence new varieties also occurred naturally.

Hybridisation means the crossing of two widely different varieties and the result is that traits from both the parent plants may be of advantage to the new hybrid variety. The hybrid will then display what is known as hybrid vigour and have a mixture of qualities that will enable it to grow more successfully than either of its parents. However, this hybrid vigour is reduced in subsequent generations.

In the seed trade this involves a deliberate narrowing of the gene pool. This means that hybrid seed needs artificial support to perform well. The result is that the genetic variability that gives plants their adaptability to local climate conditions and pests and disease is taken away.

Hybrids are bred to select for uniform ripening, for fast growth and for size. This is supported with the use of artificial fertilisers that force- feed these plants to make them grow fast. Superphosphate is used to support this accelerated growth on soil that is devoid of life and minerals. A lot of our food supply is now becoming genetically engineered to suit the requirements of the chemical companies' bank accounts. The spiel for feeding the world's poor is not valid as the third world farmer who always saved his seed for the next crop will not be able to do so anymore, as the G.M. seed has been either patented or designed to be infertile for the next generation of plants. He will then need to buy his seed every time he wants to plant a crop. It won't be cheap either. The third world farmer can't even afford the chemicals he needs to grow these seed and consequently there is a very high suicide rate with Indian farmers caught in a heavy debt cycle. This is how corrupt the whole situation is becoming. HELP! What can we do about it????

Seed Saving Networks

There are local seed saving networks all around Australia and overseas. These groups range from gardeners that save seed and exchange them with each other, to more formal groups that have seed banks where groups like Permaculture groups and organic gardening clubs are contributing, swapping, and selling seed with each other.

There are also locally grown seed for sale in produce stores and other local shops. You will do well to support them and then pass them on to friends and neighbours.

The advantage of local seed saving networks is that the seeds are sourced locally and can therefore adapt to local conditions. These seeds will generally perform well and are easy to grow. The plants grown from open-pollinated seed are often better in flavour, texture, nutrition, and tenderness.

On the other hand, most food on the supermarket shelves, including fresh produce, is grown from hybrid seed. Ever tried to grow from seed from fruit and veg from the shop? Some cereals, fruits and vegetables sold in health food shops are the result of hybrid seed that is lower in nutritional value. Many of these seeds either don't germinate or they grow plants that are throw backs from their ancestors.

Home gardeners do not need to grow hybrid seed as there is no advantage in doing so. It's up to people like you and me that are the backbone of saving our seed's genetic diversity to pass onto the next generation. If we don't do this, who will?

Science of Seed Saving

Pollination occurs in plants when pollen from the male parts of the flower is deposited on the female part of the flower. In most of our vegetables, herbs and flowers, the male and female parts are in the same flower. These are called **complete flowers.**

Exceptions to this are members of the cucurbit family and corn, where the male and female parts are on the same plant but in different flowers. These are called **monoecious.**

Asparagus have male and female plants, and these are called **dioecious.**

Self-pollination: In some complete flowers, self-pollination occurs before the flower opens. Lettuce, tomato, peas, beans, and okra have the female part of the flower so close to the male part of the flower that the slightest wind movement, even from a passing bird, causes the pollen to drop into the receptive stigma (the female part).

This is called **automatic self-pollination.**

Cross-pollination: Other types of complete flowers require cross-pollination. They need insect or the wind to create fertile seeds.

Some plants such as the brassica family have a chemical barrier that prohibits self-pollination within the flower. They require bees or other insects to bring pollen from another plant to effectively carry out cross-pollination. If one plant were grown on its own it would have hardly any seed. Pollen from other brassica varieties can easily cross with each other to form broccoli- cauliflower crosses etc.

Rogueing with cross-pollination will of course contaminate the true types.

Natural cross-pollination:

Plants such as lettuce, tomatoes, peas, and beans are self-pollinated and do not rely on insects or pollen from other plants to produce fertile seed. They are called in-breeders. However, in a garden situation, a certain amount of natural cross-pollination happens because of curious and hungry insects. Pollen often sticks to the body and legs of insects. Some pollen of the same family is more dominant in one variety than another. Eg. Chilli is more dominant than capsicum and it can therefore revert to a chili type capsicum if these are kept in the garden at the same time of flowering. Hot chilies will compete with other chili varieties, and one will dominate the other. The same is said for basil. For this reason, self-pollinators should be kept as isolated, if possible, for professional seed saving.

Keeping your seed pure – this can be done in several ways.

Grow them apart from each other to prevent contamination from insect or wind-blown pollen. How far apart differs for each plant. (See the seed saving ratings for ease of seed collecting and do a search on specific vegetables and their distance requirements.) Obstacles such as solid fences and buildings can break up the flight path of insects and will greatly reduce the chances of crossing.

Experiment with varieties that are not rare.

Isolate the plants with staggered planting times, this is the easiest method to ensure purity.

Bag the plants- when only a small amount of seed collection is needed, covering the blossoms of fruit such as tomatoes and capsicums with a paper bag or a pantyhose. This is for automatic self-pollinating plants only. Plastic bags are not suitable because they prevent airflow. The bags will exclude insects and any pollen that is flying about at the time of flowering. The bags can be removed once the fruit is set.

Biennials are plants that produce vegetative growth during the first growing season, slow down through a period of cold weather, go to seed the second growing season, and then die.

Root crops can take two calendar years for seed production.

Onions, carrots, and celeriac seed can be obtained by root to seed production. Simply plant the sprouting end of a carrot or onion and allow it to grow and form the seed head then save for seed saving. This will take only half the time as in seed-to-seed production.

Perennials are the types of plants that can survive on neglect. This is one of the main reasons why so many different varieties have been adapted into permaculture gardens.

Root crops such as sweet potatoes, taro, arrowroot, yam, ginger, turmeric, and water chestnuts, all die back in the cooler weather in the sub-tropics only to re-shoot again when the weather warms up.

Leaf crops such as aibika, New Zealand spinach and kang kong will die back with the cooler weather and re-shoot again in spring/ summer.

Legumes such as the seven-year bean varieties (there are many) yam bean, Madagascar bean, lablab and the scarlet runner bean all die back in the cooler weather only to regrow again in spring.

Other plants can be divided by crowns- such as lemon grass, rhubarb, asparagus, yakon, and divider leeks.

These plants will keep producing more planting material by vegetative propagation.

Seed Saving Simplicity Rating

For the beginner	For the gardener with experience	For the experienced seed saver	For the expert seed saver
• Basil			
• Beans	• Amaranth	• Beetroot	• Brussels sprouts
• Broad beans	• Asparagus	• Cabbage	• Corn
• Cassava	• Basella	• Cardoon	
• Chilacyote	• Bitter Gourd	• Cauliflower	
• Choko	• Borage	• Celeriac	
• Coriander	• Broccoli	• Collard	
• Dill	• Calendula	• Corn salad	
• Eschallots	• Cape Gooseberry	• Cucumber	
• Fennel	• Capsicum and chilli	• Eggplant	
• Garlic	• Carrots	• Endive	
• Garlic chives	• Celery	• Kale	
• Lemon grass	• Celtuce	• Kohl rabi	
• Lettuce	• Chervil	• Mustard	
• Lima and Madagascar bean	• Chicory	• Mustard greens	
• Marigold	• Chinese cabbage	• Onions	
• Nasturtiums	• Chives	• Orach	
• Peas	• Cow pea	• Parsnip	
• Sage	• Dandelion	• Radish	
• Salad burnet	• Ginger	• Rockmelons	
• Snake beans	• Gourd	• Silver beet	
• Sweet potatoes	• Gramma	• Spinach	
• Tomatoes	• Hibiscus spinach	• Tarragon	
• Watercress	• Hyacinth bean	• Water	
• Yam	• Korilla	• chestnuts	
	• Leek	• Water spinach	
	• Luffa		

	◆ Marjoram ◆ Mint ◆ Mitsuba ◆ Mizuna ◆ New Zealand spinach ◆ Okra ◆ Pansy and violet ◆ Parsley ◆ Peanuts ◆ Peruvian parsnip ◆ Poppies ◆ Potatoes ◆ Pumpkins ◆ Rhubarb ◆ Rocket ◆ Rosella ◆ Rosemary ◆ Runner bean ◆ Salsify ◆ Sorrel ◆ Soya bean ◆ Spring onion ◆ Squash ◆ Sunflowers ◆ Taro ◆ Thyme ◆ Tree onion ◆ Turmeric ◆ Watermelon ◆ Wax gourd ◆ Winged bean ◆ Yam bean	◆ Jerusalem artichoke	

When to pick the plants for seed saving

To make sure that the seeds have matured enough, it is important to leave the plants in the ground until the plant has died and dried up. Tomatoes and other fruits that ripen and fall to the ground need to be collected when they are over-ripe, and beans and peas should be left on the vines until they are brown and brittle. With plants that tend to shatter their seed, it is imperative to take them off the vine before this happens. Generally, the best time of the day is from around 10 am when the dew has evaporated.

Criteria for selection- you will need to observe which plants are the healthiest and most robust for seed collection. Select the plants that are drought hardy and that perform well. When saving seed for lettuce, it is important to save the seed of the last lettuce plants to bolt, as this encourages the characteristics of a longer lasting lettuce in the garden. There is so much genetic variability in open-pollinated plants, that with the characteristics that are required, we have choices that we can make by selecting certain traits.

Take corn for instance. It is important to select the first plump juicy cobs that form on the stalks. This is to encourage the early forming of the corn cobs. With beans, it's best to select a few bushes or vines that will be grown purely for seed. The first beans are especially important. When the beans are left on the vine it will slow down production, so some sacrifice needs to be made as the more beans you pick, the more productive the pod set. With root crops, choose the largest, smoothest, and most well-formed specimens that are the best representatives of that crop. It is those characteristics you continuously select that will dominate over the next few generations. Selection for certain characteristics can take up to ten years to take effect.

When selecting certain plants for seed saving, they need to be marked so that no one will touch them. This also means YOU. Tie a ribbon on the plant or place a sign nearby to identify it as a seed saving plant.

How many to select- more rather than less is always better. This is to ensure a genetic variability. Some plants preserve their genetic variability better than others and here are a few examples.

Self-pollinating plants such as tomatoes, lettuce, beans, and peas need one to six plants reserved for seed saving. They are natural in-breeders. If you only save the seed from one plant year after year, this will eventually lead to inbreeding-depression and will eventually cause the strain to run out.

Cucurbits such as pumpkins, melons and cucumbers need to have around six fruits saved for seed saving. It is an advantage to keep seed from pumpkins of different vines of the same strain rather than from one vine.

Corn and onions need seed saved from a larger number of individuals to maintain genetic variability. Fifty to one hundred cobs of corn need to be saved whilst the seed heads from about 20 onions and leeks need to be kept maintaining enough genetic variability.

Processing the seed for storage

Cleaning- you can clean the seed either by a wet or dry method.

Dry seed cleaning is the method used for seed that mature in a pod or a husk such as beans, corn, lettuce, and most garden flowers. If the rainy weather sets in before maturity occurs, the plant can be pulled up by the roots and hung upside down or laid on a table to thoroughly dry before processing the seed. Dry pods can be harvested straight off the bush.

Winnowing the seed after rubbing and rolling the chaff away from the seed, the rubbish needs to be blown or winnowed away to clean up the seed.

Screening is another way of cleaning seed. Kitchen sieves and colanders work very well, and it is good to have sieves of different gauges to adjust to a variety of seed sizes.

Wet cleaning is used for plants that carry their seed in moist flesh such as tomatoes, rock melons and cucumbers.

- Scoop the seeds out of the flesh into a large container of water and rub them vigorously.
- Collect seed with a sieve and run water over them to remove all the little bits of flesh.
- The clean seed can then be placed on a plate until thoroughly dried.

Fermentation The seeds of tomatoes, passionfruit and cucumbers can be fermented to get rid of seed-borne disease, by the action of bacteria and yeasts. This also makes the seed nice and clean for storage.

- Cut the fruit in two and remove the seeds and pulp with a spoon and add a little water into a glass jar. Leave this to sit at room temperature
- After a few days foam will appear indicating that fermentation is complete, and the surrounding gelatinous pulp has dissolved
- Clean the seeds by rinsing thoroughly with water
- Spread out on a plate to dry

Hot water treatments-This is a safe treatment for disease such as black rot, black leaf spot and black leg in cabbage and downy mildew in spinach.

- Soak the seed in water held at a constant temperature of 50 C for 20 minutes in a thermos
- After the heat treatment you can dry the seed in a sieve

Drying- this is one of the most crucial aspects of seed saving. Seed is easily ruined when it is stored while still moist. Generally larger seed will need a longer drying time than smaller ones. The way to test if a large seed is dry enough for storage is to bite into it. There should be no indentation made with the teeth.

- Dry seed in a shady place on a plate
- Hang small quantities in paper bags and hang up to dry
- Lay larger quantities on an old window screen

Diseases that are spread in, or on seeds needs to be avoided particularly if you are passing seed onto other people.

Storing seed Most vegetable and flower seed have a life span of three to five years. There are those seed like parsnip that have a very short viable seed life (they generally don't last longer than from one season to the next. The entire umbelliferous family and allium genus have rather short-lived seed as well. The viability rate of onion seed is reduced by more than half if the seeds are stored in a warm place such as a west facing room in summer. Thick- coated seed tend to last longer and larger seed also tend to stay more viable than the smaller varieties.

Seeds in storage are dormant but still alive. To maintain seed viability, it is important to allow only a minimum exchange of gasses by keeping a constant temperature and humidity rate. In the open air the seed will absorb moisture and the nutrient stored inside the seed will start to oxidise. When the temperature rises, the seeds will release carbon dioxide and generate more heat. Soon the respiration rate will rise to an unacceptable level and the seed won't be viable anymore.

Darkness the action of light also has a detrimental effect on seed viability. Put them in paper bags or dark coloured jars or store them in a cupboard.

Moisture excessive moisture will allow the seed to overheat and become like compost. Different seed coat thicknesses will determine how soon moisture will affect seed viability. Most seed need to be stored at 10% or below humidity. (Peanuts and soybeans are best stored at 15% because of their oil content.

At lower moisture levels seed can handle temperature variations better.

Silica gel crystals can be bought at chemist stores and will absorb any moisture from the seed in a sealed container. The colour of the crystals will indicate how much moisture has been absorbed. Blue is dry, pink is moist. Dry the crystals in the oven until they turn blue again. Dry rice grains or salt will do the same job. You can also save silica crystals from shoe boxes and vitamin pills.

Some exceptions avocados, malabar spinach and pawpaw seed should not be dried before planting. Rain forest seed should be kept moist and cool, and these can be stored in a plastic bag with some coconut coir fibre until they are ready for planting. For most vegetable seed, storage temperature of 5 C is ideal and for long-term storage, a fridge is the ideal place. For short-term storage, under the house in a south facing room would be fine.

Insect damage-Two days of freezing the seed won't hurt but it is important to make sure that the seed are thoroughly dry. Place the seed into a container before freezing and don't open the lid until the container is back to room temperature. Freezing will kill any living creatures such as weevils.

Seeds that are at the end of their potency are often weaker and produce plants and seeds that have genetic defects or show a lack of vigour when the plant is growing.

Practical Seed Saving Around the House and Garden

Selecting plants for seed saving

- have ribbons, strings, labels etc handy for earmarking plants
- pen or pencil for writing on the labels

Where around the garden are these kept? _____

Collecting seed for processing
Have secateurs, knife or cutting implant handy harvesting

Where are these kept? _____

Have bags handy for transporting seed heads in from the garden

Where are these kept? _____

Have an area where you can store seed heads until ready for cleaning, drying, and packaging

Where is this situated? _____

Seed processing area for cleaning and packaging

Where will this be done? _____

Raising seedlings

There are many factors to consider when planting out seed for successful germination

- Soil temperature
- Air
- Suitable time of year
- Depth and spacing
- Available sunlight
- Even soil moisture
- Day length hours
- Viable seed
- Chilling of some seed (stratification)
- Scarifying (scratching the seed coat)
- Smoking and fire for some native seed
- Hot water soaking
- Keep out vermin and other animals

(Farmers have observed that three-year-old pumpkin seeds will produce more female flowers than fresh seed)

How a Seed Germinates

- Once the seed is planted it simply absorbs water and swells. Seed will absorb around 50% of its own body weight in water as they begin to germinate, and they must not dry out at this stage
- The seed coat will then crack open, and the seed begins to breathe
- Uptake of water continues and the reserves of the starches within the seed are activated for food
- Cell division begins and the root radicle emerges
- The leaves then push up towards the light
- The seedlings can now have a light feed of aerated compost tea or vermicast liquid

Most plants need darkness to germinate but some need sunlight

Germination tips

Problems with the growing medium

- May be too wet- too much water causing seed to rot
- Maybe too dry- cover seed with some shade-cloth or create a humid tent with a plastic covering making sure there is adequate air circulation
- Fungal disease- use fresh seed as much as possible
- Growing medium too well-drained or compacted

- Planting depth, very fine seeds need to be pressed onto the soil mix. Lower soil temperatures will adversely affect seed planted too deep as the seed can rot. In a hot climate however, corn seed may be planted 5cm deep to prevent drying out and give the developing roots some soil depth to prevent them from falling over in windy weather
- Keep out animals to prevent them from digging up seed.

Seed raising mixtures

One part of well-prepared compost or vermicast, one part coco peat and one part of vermiculite. Some sharp sand can be added as well. Other suggestions are as follows:

- Add one part of vermiculite/ perlite to two parts of sharp sand
- Use 50:50 vermicast with sharp sand
  Fertiliser should not be added to seed raising mixtures as the seeds themselves do not use fertiliser to germinate. The young emerging roots may get burnt with it. (Vermicast and well-made compost do not burn tiny rootlets so they are a very safe and beneficial ingredient to use in potting mixes)
- Vermiculite is good as it can absorb large amounts of water and nutrients Vermiculite is expanded mica and the tiny root particles can penetrate its particles
- Perlite is of volcanic origin and is mined from lava flows. It has no nutrients
- Garden soil is fraught with difficulties such as soil-borne disease and compaction in containers
- Water retentive gels or granules can be used as well

A good potting mix should have a balanced pH, should be well aerated, drains well and is easy to re-wet after drying out.

Filling the pots for seeding out

Hygiene is important and used pots should be washed prior to being used. They can have a final rinse in 1% bleach or a good shot of vinegar.

Pots need to be filled to the top and tapped onto the bench top to exclude any air pockets. Punnets should be banned as these are far too shallow for the home gardener. The potting media dries out far too quickly without a sophisticated watering system and there's little room for the roots to grow. A four-inch pot is ideal as it gives depth for root formation but doesn't take up a lot of potting media. Twenty pots can fit into a tray for easy handling.

The mixture should be neither too wet nor too dry but nicely damp. Do not over water or under water your seedlings. Feed your seedlings with a weak tea of kelp/ fish emulsion or diluted worm juice once a fortnight. Make sure that the emerging seedlings have enough sunlight. More sun is needed in the winter than in the summer. The morning sun in summer is kinder than the western afternoon sun. Keep the sunlight filtered so that the young seedlings do not wilt during the day.

Propagating Plants

Many perennials can be divided by root division. A lot of our Permaculture plants fall into this category and hence it is very easy to build up a good stock of planting material for your gardens.

Growing from cuttings

Many plants can be propagated with cuttings. The best time of the year for this is when the temperatures are from 20- 25 C. Cuttings do best from shrubs and trees that have been regularly pruned.

Cuttings fall into three main groups- stem, leaf, and root cuttings. Stem cuttings fall into three sub-groups, and these are softwood, semi- hardwood, and hardwood cuttings.

Tools and equipment

A good quality knife, a sharp pair of secateurs and a razor with one edge covered with insulation tape for easy handling, are very useful tools for cutting. A plastic bag to carry the cuttings in transit is also a good idea as it's most important to always keep them moist, especially for softwood cuttings. Make woody cuttings as thick as your thumb.

Softwood cuttings

These can be taken from shrubby plants with soft green shoots. These cuttings are in an active growth stage and no wilting is allowed. Take these types of cuttings in the early part of the morning when the plant is full of sap. There are however some exceptions to this rule, and these are the geranium type plants that need to dry out for a day or so before potting on. This is because they need to form a callous to prevent rot from occurring. The best time to take cuttings is when the moon is waning to put more energy into root formation.

With tip- cuttings pull the surrounding leaves off around the area where the stem is to be cut. Make a clean cut at a slight angle just below a node or leaf axil. For plants with large leaves, take all the leaves off except for the top couple. These can be cut halfway across just so there is some leaf available for photosynthesis. Tip-cuttings should be about 5- 10cm long and have four to six nodes. Heel cuttings are also used in the case of conifers and many other plants.

Semi-hardwood cuttings can be from woody pieces of branches where the woody part is transforming from green to a firmer brown or grey wood. These can be from side-shoots or a lower part of the more mature wooded sections. Again, take of all the leaves except for the top two and cut these in half.

Hardwood cuttings are taken from deciduous trees when they are dormant. Choose wood that is up to 20mm in diameter for cuttings and cut the sections into 20 cm lengths. A handy hint is when collecting cuttings, cut the bottom end at an angle and the top end straight across. This will remind you which way is up. Plant just before budding starts at the end of winter.

Containers and potting mixes-

All containers must have free drainage and the potting mix must be porous and yet able to hold moisture as well as be free from disease.

A mixture containing two parts of coarse sand, one part vermiculite or coco peat and one part charcoal is very suitable for striking roots on cuttings. There is usually no fertiliser present in this potting mix, but adding phosphorus is useful to encourage rooting.

Root-promoting hormones (or cuttings dipped in honey) can help cuttings to grow roots quickly. Dip the cut ends into the powder before planting and with a pencil, make a hole into the potting media and carefully insert the tip cutting into it.

Micro-climate is important especially if the surrounding air has low humidity. Placing a plastic covering over the pot or container will help to create a humid atmosphere. Do allow some free-flowing air to circulate otherwise there can be a mould problem. Don't expose the cuttings to full sun but rather choose a semi-shaded area.

Layering and marcotting are two methods of encouraging living root stock to form roots while still attached to the plant.

Layering is the easiest method of the two and it is simply a matter of drawing down a lower branch of the plant and pushing a tent peg over it and making a forced contact with the ground. Many plants will effectively start to strike rootlets (like strawberry runners) but some woodier plants might need a cut made into the underside of the branch so that when the branch is pulled down it will open up slightly to expose some of the inner part of the branch. The rootlets should begin to form from this cut.

Marcotting is a similar method but more suitable for higher up in a tree. To reproduce a particular tree, you need to have a plastic bag with a soil mix in it. Choose a branch that is no more than a large thumb size in thickness and with a very sharp knife work away one inch of the bark all around the circumference. Choose soil that has some clay content in it to allow it to stick together somewhat and place in the plastic bag, pack this against the cut section of the branch and wrap the plastic tight around it.

This will work when the tree is in the active growth stage especially when the weather warms up and in a period with high rainfall to ensure plenty of humidity. Within a few months you will have a tree large enough to plant out and have a great head start, and it's free!

NB: it helps to sow seeds and take cuttings in the right moon phase. From new moon to first quarter has the most vigorous germination rate and the third quarter moon is best for taking cuttings. It's very worthwhile to study moon planting for best results.

Build a nursery

The Permaculture nursery- quick and cheap made from stakes, poly-pipe, and shade-cloth.

For a small nursery-

- Six star pickets
- Poly pipe 2" wide (to be able to slide over the star pickets)
- Shade cloth
- Small sections of poly pipe to turn into clips to hold the shade cloth in place
- Metal door jamb and screen door
- Two lengths of reinforcement rods and bolts to connect to poly pipe

Inside the nursery-

- A potting bench
- Trough for washing pots
- Storage area for pots, trays, and tools
- A white board for notes and reminders / corkboard to attach notes
- A general-purpose bench for hardening off plants

Aspect of the nursery should be full sun most of the day
Shade cloth from 50 to 70% shade
The benches should be positioned in full sun and be made from metal as wood can harbour fungal diseases
You will need a good supply of pots and trays within easy reach and extra benches with roller coaster wheels could be useful
An ironing board can be used as a potting table
It's best to pot up outside of the nursery where there is more fresh air
The floor should be covered with fine gravel or weed mat for easy drainage
Place the potting area on the south side of the nursery away from the hot sun

A seedling raising nursery with a plastic covering over it can be very useful when humidity is very low. These can be bought at various hardware stores as they are made from a metal frame with shelves and a plastic zip up covering to create a humid atmosphere. Once the seedlings have germinated the plastic covering can be replaced by a shade cloth cover, otherwise it can get too humid for the young tender seedlings.

What to plant when

For Southeast Queensland and Northern NSW
Harvesting is usually around three to four months from planting the seedlings.

March / April is the time to start planting temperate vegetables. July/ August is usually the last month for planting these vegetables except for beans, zucchini, Asian greens, celery, beetroot, tomatoes, carrots, corn, and all cucurbits such as cucumber and pumpkin and lettuce until Christmas.

Plant from March/ April

- All brassicas (broccoli, cauliflower, cabbages) including the Asian greens and cabbage varieties (such as bok choy, pak choy etc.)
- Onions, leek, and garlic no later than end April. Harvest onions and garlic when tops have died down in November/December
- Potatoes
- Carrots and Parsnips
- Kohl rabi swedes and other turnips
- All leafy greens such as lettuce, spinach, silver beet and mustard greens
- Cucumber, zucchini, and pumpkin especially if in frost free areas
- Tomatoes, any variety if in frost free areas
- Bush beans and peas
- Rhubarb
- Celery
- Beetroot
- Broad beans if planted from February/ March

Some of the temperate vegetables do well in our hot summers especially if the moisture is kept up to them.

Plant from September:

- Rock melon, watermelon, cantaloupes, and all other cucurbits.
- Corn
- Beans
- Sugar loaf cabbage (if there's no white cabbage moth)
- Capsicum and chilies
- Egg- plant
- Okra
- Rosella
- Pak choy and bok choy

There are also some rather unusual vegetables:

African horned cucumber- sow in spring
Peeled and sliced as thinly as possible. Cover with yoghurt and chopped mint. Puree for a cold summer soup

Asparagus- can be planted from seed but growing from crowns will be faster. Plant in spring in trenches enriched with plenty of rotted cow manure and organic matter Asparagus plants will last up to twenty years if regularly fertilised and kept weed-free.

Buckwheat- cover crop
Can be sown onto poor ground to make phosphorus more available- sow all year round but does best in the cooler time of the year

Cape gooseberry- spring, perennial in our climate

Celeriac- plant the same as celery- mainly throughout the cooler months but can also be planted in spring

Collard- very hardy but needs adequate moisture. Plant in autumn to spring. Eat the leaves fried in bacon fat

Divider leeks- plant out from parent plants in March- April

Ginger- plant in September to November and plant 5- 10cm deep Needs a very fertile soil with even soil moisture

Horseradish- Plant in spring and harvest in the late autumn when the leaves die down Plant will grow back the next spring if the garden soil is moist enough

Kale is the ancestor of cabbage- plant late summer to autumn
Buttered kale with sauteed garlic and lime juice

Millet is a fast- growing annual summer grass. Grows well in poor soil. Use as a cover crop. Needs regular rain fall. Can be used as a fermented porridge Ogi

Strawberries- plant after the runners have set in autumn to spring. Plant around garden areas and next to footpaths. Replant every 3 years to encourage berries to grow

Sugar cane- is grown mainly for mulch. The black variety is very popular. The sugar cane juice is very nutritious and makes a great drink for kids. Let them chew on the slivered cane stalks as a simple treat

Vegetable spaghetti- is a squash variety. Plant in spring. Same requirements as for zucchini

Welsh bunching onions- grow by division. Sow in spring or divide any time.

All season planting beans, beetroot, herbs, lettuce - non- heartening varieties in summer, mustard and cress, radish, shallots, and tomatoes except for frost

Tropicals:

Plant all tropical vegetables from spring onwards

Plant division- many perennials can be multiplied by root division. A lot of our Permaculture plants fall into this category and hence it is very easy to build up a good stock of planting material for your gardens

Irrigation

Overhead and drip irrigation

The two most widely used irrigation systems used are overhead and drip or trickle.

Overhead irrigation is designed to cover a large area, and these systems are the least expensive to install. However, this method produces uneven water distribution, which can slow plant growth, encourage disease, and contribute to runoff.

Large containers are usually watered with a drip or trickle system, which uses 60%-70% less water than overhead systems. Drip irrigation systems cost more to install than overhead systems, but they have superior application uniformity and efficiency. They are also less affected by wind and crop canopies, and they produce less runoff. Another advantage is that you can continue working while the plants are being irrigated. The biggest disadvantage to trickle irrigation, besides the initial cost, is keeping the pipes and emitters clean.

A third, ever more popular watering method is wicking beds. This method has been described in the different types of gardens section in this book.

Using grey water

Greywater is the wastewater from your washing machine, shower, bath, and basins.

There are two types of greywater systems, each allowing you to use greywater in different ways.

- **A greywater diversion device** enables untreated greywater to be used for outdoor purposes by distributing water to your garden through a sub surface irrigation system.
- **A greywater treatment system** enables you to use treated greywater for above surface irrigation. Treated greywater can also be stored.

For more information about the correct use of greywater do a search on your local council website.

Things you should know about greywater

Greywater is not permitted to run off your property.

The Australian Standard requires that a sign be placed on the outlet of the greywater diversion device marked "WARNING DO NOT DRINK" and that all irrigation pipes be coloured purple. Be extra careful on what goes down your sink outlets. Be very diligent to use only garden safe products for sub-surface irrigation systems. I have been using laundry balls for many years instead of using washing machine products. Sub-surface irrigation systems work best to distribute greywater evenly around your garden.

Prepare grey water for irrigation

- Wastewater directly from the bath or washing machine should not be simply directed through diversion devices without at least filtration. Bath or shower water may contain hair. Laundry water will contain lint. Simple filtration can be provided by place an old nylon stocking over the end of the pipe. Water from rinsing, soaking, or preparing nappies for wash should be discharged to the blackwater system and not surface spread on the garden. If chlorine bleach is used to soak nappies, or some of the 'oxygen rich' products, don't dispose of that water onto your lawn or garden. These disinfectants are likely to have detrimental effects on the soil microbes that are essential for nutrient cycling.

Local rules

If all Australians were the same, we could expect to have the same rules and regulations regarding greywater reuse on our suburban or urban lot. It appears we are not the same because there are different rules for not only different states but also in different local government areas.

- If you are prepared to bucket out your greywater, then anything goes, anywhere, anytime, any amount.
- But if you would like to plumb a system, be aware of the local rules

How long and when to water

- In sandy soils you can usually apply a lot of water quickly and it will be easily absorbed.
- In heavy clay soils you should water slowly over a long period. Heavy applications will not soak in quickly enough, and a lot will be lost as run off.
- Deep rooted plants such as trees should be watered slowly over a long period, to wet the soil to a greater depth.
- Deep rooted plants can be watered less often than shallow rooted ones. Shallow rooted plants such as annual flowers and vegetables need frequent watering, but of a shorter duration at each watering.
- Consider the time of day when you are watering. It is best to water when the plants will gain the most benefit and won't lose a lot of water through evaporation and transpiration. This is normally worse when it is windy, hot, dry, and sunny. So, if these conditions apply in full or part, try to water early in the morning or/and evening.
- Just because the garden looks wet doesn't mean that the water has reached the plant roots. Take a stick and dig under the soil in your garden to see how deep the water has penetrated. Infrequent deep watering is far preferable to frequent light shallow watering as it encourages the roots to expand downwards and outwards seeking the deeper water in the sub-soil.

Water restrictions for irrigation

When gardening with town water there will be times when there are water restrictions. There are probably restrictions on set times but if you are allowed to water then:

- Sprinkler will use 111 buckets of water per hour.
- Watering your garden by hose uses 600 to 900 litres of water per hour.

- Clearly an enormous volume of water is used by most gardeners to maintain a healthy garden. The problem is that over 80% of this water can be wasted, because of water runs off to areas where it is not needed by plants and through evaporation.

A water fork or watering probe will deliver water straight to the roots thus avoiding surface evaporation. This is capillary action and will encourage deep rooting in plants.

Most watering only wets the surface layer of soil, and this process promotes shallow roots and spindly plant growth. By injecting water to the roots of your plants, not only does this save you water, it also encourages stronger, deeper root growth and healthier plants.

Additional benefits of the watering probe are, it waters the roots of your plants so you will now be able to:

- grow plants on slopes where surface watering would normally run off without wetting the soil.
- grow plants in paving where there is not enough soil surface area for an adequate supply of water to penetrate
- reduce black spot by watering the roots, not the foliage of plants such as roses, tomatoes, azaleas etc.

A watering probe is also great for getting water into compost heaps.

How often do you need to water?

Nothing is more important for your plants ability to thrive than having enough water. Adequate watering would save the most plants that are lost in gardens each year. The best way to tell whether your garden needs water is to feel the soil 2 to 3 inches (5 to 8 cm) below the surface. How often you should water will depend on factors such as the type of soil, how well the garden is mulched, and the amount of sunlight your plants are exposed to. Assuming adequate mulching and summer temperatures are up to mid-20s C then watering once a week should be enough. However, with full sun and temperatures in the 30s C then you may need to water every second day.

Don't forget that a thorough watering around the roots of your plants with a watering probe once a week is far more beneficial than spraying the soil surface every day.

Plants need the soil to be at least 50% moist for nutrient uptake.

Water bags for around trees act like a mulch at the same time.

The ECO BAG is an automatic drip watering system costing considerably less than replacing trees and shrubs which fail through inadequate watering. This unique system consists of a plastic bag that drips water into the roots of a plant or tree over a long period. It has a long life and will establish more than just one tree. It has been developed in dry areas and is now fully proven after an exhausting three-year period, to provide an adequate controlled water supply to establish young trees.

- Eliminates evaporation from around the root zone while still allowing rain to enter. Automatically waters trees and shrubs

- Eco bags provide the perfect insulation around the root zone, warming in the cool weather and cooling in the hot weather
- Enables trees and shrubs to be planted at any time of the year and to grow through summer
- Saves the enormous costs of plumbing and irrigation involved in orchard establishment.
- Holds approximately 20 litres of water, lasting from 3 to 4 weeks. Can be easily refilled
- The flow of liquid can be varied to suit the size of the plant or length of time water is required to last
- Provides for years, the perfect grass and weed control around plants. Allows the addition of fertilizer
- Provides excellent support for small trees and shrubs
- Easily hidden, if desired, under mulch (wood chips etc.)

Benefits of drip irrigation Drip irrigation is the most efficient watering method around, proving that low flow is the way to go. Hardly any water is wasted through wind, evaporation, run-off, or overspray.

Did you know that it can take 15 minutes to deliver just one litre of water to the soil with drip irrigation, compared to five seconds when delivering one litre by hand?

By slowly dripping water into the soil at the base of plants, water is released at a rate that's easy to absorb where it's needed.

Drip irrigation also reduces the risk of erosion and soil compaction

- Uses flexible piping that is easily laid and suitable for most environments, including ornamental gardens
- Is ideal for windy areas as the water goes directly to plants roots
- Is great for everything from plants on slopes to vegetable gardens
- Will suit all soil types
- Suits all plant types
- Reduces the risk of plant fungal and insect problems (only wets the soil around the plants roots, not the leaves)
- Loses the least amount of water from wind, evaporation, and run-off
- Can be used to apply liquid fertiliser to gardens
- Doesn't waste water on weeds

Strategies for improving soil

- Increasing soil depth by mounding, or raised beds and tank gardens
- Avoiding soil tillage
- Encouraging root depth
- Increasing the porosity of the soil by the incorporation of organic matter
- Avoid rapid drying and wetting of the soil

Practices to avoid

- Compaction
- Clay contamination from sub-soil
- Powdering by excessive cultivation

Practices to use

- Capillary irrigation- where the water is drawn up from the roots
- Use lime as it reduces acidity therefore releases more nutrients into the soil
- Add organic matter
- Use cover crops

When the soil is open, friable, and full of organic matter then the moisture can be easily absorbed and kept there by the humus in the soil. Not much watering is needed if there is a light mulch covering on top of the soil. The soil should never be left bare as it is vulnerable to erosion and weed invasion.

Green manure

Green manure helps to build soil. They can provide outstanding benefits for the soil and crops.

- Increasing organic matter, earthworms, and beneficial micro-organisms
- Increasing the soil's available nitrogen and moisture retention
- Stabilising the soil to prevent erosion
- Brings sub-soil minerals to the surface and breaks up hardpans
- Improving water, root, and air penetration in the soil
- Growing of leguminous green manure crops increases nitrogen availability in the soil
- Green manure crops act as a reservoir of nutrients. These nutrients are released in the soil when they slashed down and left as mulch on top of the soil

What plants to grow
Winter:

Broad beans, tick beans, fenugreek, lupins, oats, sub clover, woolly pod vetch, yellow and black mustard seed, other brassicas, oats, wheat, or barley

Summer:

Buckwheat, cowpea, French white millet, Japanese millet, lablab, marigolds, mung bean and soybean Seeds can only germinate on bare soil and not where grass and weeds grow

What is a legume?

An important advantage of legumes is their unusual ability to obtain nitrogen, a major element needed for plant growth, from the air, as most plants are unable to do this. They do this by forming a symbiotic relationship with a group of bacteria called rhizobium, which live within a specialised structure, called a nodule, on the plant's roots. The rhizobia can take nitrogen (N2) from the air and convert it to the form plants normally obtain from the soil. This process is called nitrogen fixation.

Legumes, like lucerne, clover, beans, and peas, can fix nitrogen and will make it available to whatever follows the green manure crop. Nitrogen fixation will only occur when the right rhizobia are present in the soil. This can be checked by inspecting the roots of the plants including young seedlings. If there are no root nodules present, then you can purchase inoculum. There are three basic forms of commercial inocula, and they are solid, liquid, and freeze-dried. The most used are solid, peat-based inoculants that can be purchased for seed or direct soil application. Liquid inoculants are available in broth culture or as frozen concentrate. There are up to 17 different types of rhizobia for different types of legumes.

For maximum benefit you will do well to slash down your crop once it starts to flower. When the crop has flowered and set seed, the nitrogen content decreases. Allow four to six weeks after you dig the crop in before planting new seeds in the bed or you can use your green manure crop as a surface mulch and plant sooner. You can trial growing a cover crop by using seeds from bulk food stores that sell grains and legumes. It won't always work for you, but most seeds should germinate. These can be trialled first for germination by placing some seeds on wet tissue paper for a few days or soaking them in water and sprouting them in a jar turned upside down with a muslin cover.

Mulch

Keep adding that organic matter!

- Improves soil conditions; binding sands and opens clay soils
- Improves soil drainage
- Keeps soil temperatures cooler during the day and warm at night
- Protects plants from frost injury
- Stops erosion
- Allows the soil to be worked earlier in the spring
- Saves time and energy cultivating the soil
- Prevents surface crusting thus allowing the soil to breathe
- Reduces soil compaction
- Holds down weeds
- Prevents a hardpan being created in the earth
- Provides nutrients, gasses, and other growth substances
- Prevents vitamin loss in plants
- Encourages nutrients to be taken up by the roots
- Stops nutrients from being leached from the soil
- Reduces losses caused by soil borne diseases
- Encourages earthworms and other micro-organisms
- Encourages roots to penetrate deeper in search of food
- Helps to prevent plants from wilting
- Improves the flavour and keeping quality of the harvest
- Protects the produce from mud-splash
- Recycles wastes
- Improves the appearance of the garden
- Keeps the soil microbes happy by keeping in the moisture

If after following all the good practices you strike a problem with pests and disease, then the disease triangle can be a good tool to use as a checklist.

Disease Triangle

Host- Disease-Pathogen

Cultural practices

Cleaning up and destroy diseased plants by cooking in black plastic bags in the hot sun
Crop rotation
Control of weeds in and around the crop
Control of insect vectors
Use sterile potting media
Using disease free planting material
Monitoring, washing, disinfecting tools, and potting benches

Environment

Plant in the right season
Optimize soil pH
Spacing between plants for airflow
Soil drainage
Appropriate irrigation methods
Organic matter in soil suppresses disease

Host

Select disease resistant varieties such as open pollinated seeds
Grafting susceptible scions onto hardy, vigorous rootstock

Diseases can be pathogenic or non-pathogenic
Causes of nonpathogenic diseases could be nutritional imbalances, environmental conditions or water quality and toxins
Fungi invade host plants through natural openings on the plant such as the stomata and through wounds
The hyphae of some fungi grow on the surface of plant tissue eg.powdery mildew whereas other fungi grow inside the plant tissue such as downy mildew
Fungi spores may be spread by wind, water, seed, soil, animals, and infected plant material

But most of all keep adding organic matter.
Happy Gardening!

assembling a tyre pond

balcony garden zone 0

bath tub wormfarm

brassica crop grown in pig pen

chooks in deep litter

any organic matter is suitable for a deep litter

chop and drop lemongrass

garden produce

grey water garden with sub surface irrigation

wet seed saving

herb spiral under construction

completed herb spiral

movable pig tractor

Piggies digging the soil

planting into mulch

preparing aerated compost tea

raising seedlings in a portable nursery

seed saved by students

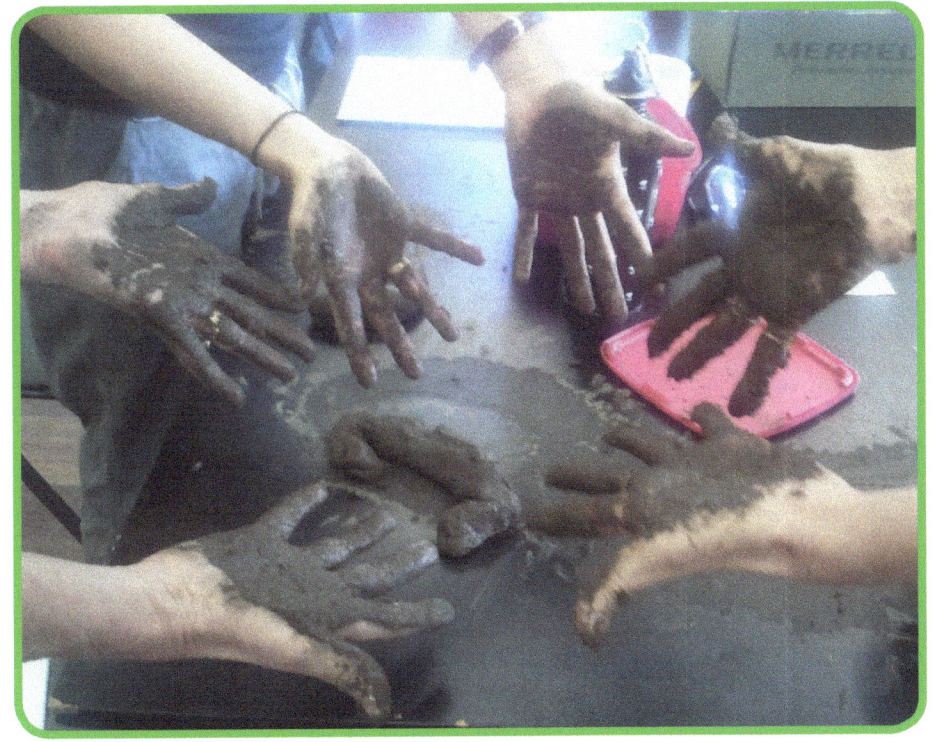

soil testing

students building a bamboo teepee

www.ingramcontent.com/pod-product-compliance
Lightning Source LLC
Chambersburg PA
CBHW041124120626
46547CB00019B/2839